Biography of
LEE MAN TAT
—Legend of the Sauce King

Kok Chung ONG
Victor ZHENG
Siu-lun WONG

The late Professor Jao Tsung-I, a distinguished sinologist and treasured scholar of Chinese art and literature, kindly inscribed the Chinese title of this book shortly before he passed away. Mr. Lee Man Tat expresses his heartfelt gratitude to Professor Jao for his thoughtfulness and deep regret for his decease.

李文達傳

遜堂

FOREWORD I

BY LEE CHACK FAN

Eighty-eight-year-old Mr. Lee Man Tat is a model of Hong Kong's veteran entrepreneurs. This biography reveals, over the last few decades, under Mr. Lee's leadership, Lee Kum Kee has overcome numerous challenges, difficulties and crises, successfully morphed from a small local business into a well-managed multinational enterprise. The biography details the legend of this discerning and determined entrepreneur, highlights his characteristics of leadership featuring "Seeing the big scene from the small picture", and sensing opportunities during crises. The former reveals his level-headedness and wisdom, while the latter reflects his composure and perseverance.

Entrepreneurs from the same generation of Mr. Lee Man Tat were in general born and grew up in misery and hardship. As we all know, our modern society has undergone great changes, as many traditional enterprises are on the decline due to a drastic shift of the market. However, Lee Kum Kee is able to keep abreast of the times with its effective transformation and incessant advancement to reach new heights. Such an achievement is certainly associated with the leader's composure, sagacity, perseverance, and diligence, as well as his aggressive spirit.

Seeing the big picture is an ability to see the world in a grain of sand. This biography is akin to a microcosm of all the hardships that the Chinese people have gone through in the last few decades. As the saying goes, "Trials and tribulations rejuvenate a nation"; likewise, success could reach Lee Kum Kee today only because it had itself overcoming numerous difficulties and crises under the stewardship of Mr. Lee Man Tat, reminding me of a little story as follows:

A wheat farmer sincerely prayed to God for a good harvest in the coming year, hoping that all the seeds planted could grow smoothly without any damages from frost and pests. God acceded to his request. Sure enough, in the following year, the farmer's crop did not suffer from frost, pests, or any other natural disaster, and he was delighted with the perfect growth of the crop. When the time for Autumn harvest arrived, the farmer gathered and threshed his wheat, yet discovered he had only collected the empty chaff without any wheat grain. Extremely disappointed, he turned to God again. God looked at the empty chaff and soon realised what went wrong. As He explained to the farmer: the growth of wheat is a natural process, of which damage from frost and pests are integral parts. The growth of plants would be incomplete in the natural world without these challenges, and hence, they bear imperfect fruits.

This is, of course, a metaphor. However, challenges, difficulties, and crises are simply inevitable in our lives. If we can face these crises steadfastly and overcome them, we will become stronger, wiser, and more successful in the future. Mr. Liang Qichao, a fellow county sage of Mr. Lee Man Tat, once expressed the same old moral as "Tribulation is the best university".

Another virtue we should reflectively learn from Mr. Lee Man Tat is the core value he has been embedding in Lee Kum Kee, which is also his personal motto: "Si Li Ji Ren" (considering others' interests). Enterprises of any types will certainly keep concerning their profits in mind, yet at the same time, they should take others' interests into account as well, in particular, the interests of their clients and the general public. Such a conviction of Mr. Lee is highly consistent with the traditional Chinese virtue of "Seeking betterment for the people worldwide" or the Bodhisattva spirit of the Mahayana Buddhism in "Benefiting oneself through altruism". This spirit is definitely conducive to the long-term sustainable development, continued expansion, and pioneering of Lee Kum Kee. With this belief in mind, Mr. Lee is always happy to contribute to charitable causes and society's needs.

I appreciate the efforts taken by the three scholars to write such a good book for the public. It is indeed a superb textbook that imparts on our younger generation the wisdom they need in order to initiate their route to success. Instead of using abstract theories to preach, the authors choose to narrate the story through a series of vivid real life incidents. I sincerely hope that our young students can thoroughly read through this fine book, as they will surely be inspired and enlightened, and therefore benefit very much from it. Our education system today attaches great importance to the transfer of knowledge, but not much to the wisdom of life. This book is really the best supplement.

Lee Chack Fan
Former Pro-Vice-Chancellor of the University of Hong Kong,
Academician of the Chinese Academy of Engineering
December 2017

FOREWORD II

BY LEE MAN TAT

The world is an ever changing place fraught with uncertainty. It brings risks and pressures as well as opportunities and delightful surprises to mankind. These conflicting elements usually appear at the same time and intertwine with one another, thus, entailing wisdom to identify and transform them.

On the other hand, the world is holistic in which all things on Earth are intricately related to one another and could appear as confused and obscure. What we can grasp may just be the tip of the iceberg, but so long as we are able to see the world in a grain of sand and discover the unspoken rules of functioning, we may decode the obscurity and strive to advance from small to large.

In 1888, my grandfather, Lee Kum Sheung, invented oyster sauce and created the Lee Kum Kee brand. This was also the inception of the Lee Kum Kee family history, which has spanned 130 years. As the third generation of the Lee Kum Kee family, I take it upon myself to perpetuate our heritage through future generations. I have experienced the ups and downs of my family's development and actively propelled the rapid advancement of Lee Kum Kee. I officially took the helm of this family business in 1972 when it was just a small firm. However, after more than four decades of toil and sweat, Lee Kum Kee has undergone a dramatic transformation today.

However, if I did achieve a few things in my life, it must be the outcome of luck, and a certain amount of wisdom.

I am very fortunate to be part of this era and of the Lee Kum Kee family. I was particularly lucky to meet the love of my life and have my five children with her. And last but not least, I was extremely blessed to possess the wisdom of how to get along with people and deal with matters when I was very young. Such wisdom allows me to see the big picture from the small detail and to grasp opportunities in crises; as it also enables me to choose the right things, to accomplish my goals, and to achieve something most significant.

An old saying that may sum up this wisdom is "Always consider others' interests in whatever we do". In other words, one should ensure the sharing of interests amongst all the stakeholders in business instead of monopolising or maximising the interest of one's own.

Indeed, such wisdom equips us to discern things that can really "benefit all of us" in the face of changes, and find out ways to get the job done confidently and seriously. In times of crises, we may also be able to "remain unflappable under pressure and bear a conviction that sages may still be around even today". With this confidence, we can turn the tide by seizing opportunities and then confidently move forward.

Lee Kum Kee was started humbly as a small workshop producing oyster sauce, and has striven in the field of condiment for over a century now. To casual observers, it is an industry with a very low threshold to start up, thus, everyone can try and then have a share of the pie. However, a condiment is indispensable for our daily meals, and therefore, closely related to our lives; every portion or every drop of it needs to be strictly assured with quality. It is precisely such a pursuit for the best quality that shapes the present-day Lee Kum Kee. Accordingly, Lee Kum Kee's condiment has also been able to reach outer space as it becomes part of the food packages of astronauts. As a basic commodity, a condiment not only implies an incessant need in the market, but also a representative of the nation's culinary culture.

Twenty-five years ago, Lee Kum Kee began its second business in health products based on Chinese herbal medicine. Likewise, it is closely associated with people's lives, and it has kept up with the development of the times. Apart from its vitality, the new business is also a representative of the nation's health regimen.

These two businesses of Lee Kum Kee are particularly meaningful and have plenty of scope for development. Promoting Chinese culinary culture and health regimen means that we are catering the needs of health and happiness worldwide with the quintessence of Chinese culture that has spanned five thousand years. After all, to keep a healthy and happy life through our food and health regimen is the basic need of mankind. It is only with such understanding that Lee Kum Kee is able to survive a hundred years of vicissitude by overcoming numerous difficulties and hardships.

I set out to work when I was still a teenager, and I have been engaged in this business for more than seven decades now. Throughout this long journey, I have experienced every joy and sorrow of life and suffered through politico-economic turmoil. During the course, I have read the richest tapestry of life and have been enlightened by the wisdom that has precipitated it. My life is full of imprints of the times, the epitome of the development of Lee Kum Kee, the indulgence of my family, the support from my friends, the life-long pursuit for a better future, as well as the regularity and wisdom of life. All these have contributed to my serenity in the face of life's ups and downs. The possession of such wisdom enables me to distinguish between the risks and opportunities.

I thank the three scholars for turning my life into a series of intriguing stories through their professional expertise and unique perspective, with contributions from my family, friends, business associates, and colleagues. They have ingeniously integrated my thoughts of life into these stories, animating my life trajectory to be shared with the readers. They have, in fact, fulfilled a wish of mine that I've had for many a year.

I would also like to take this opportunity to thank my grandparents, my parents, my wife, my children, and grandchildren, my business partners, friends and colleagues. They are the ones that inspire me and enrich my life.

PREFACE

The World Is My Oyster:
Grasping Opportunities In Crises Through "Seeing The Big From
The Small"

On July 28, 2017, the media around the world extensively reported that
the Lee Kum Kee Health Products Group had acquired the landmark
commercial building in London — also known as the "Walkie Talkie"
building — at No. 20, Fenchurch Street, for about HK$12.8 billion. It was
a sensational item of news worldwide for two reasons. First, it surprised
many to see the acquisition of a top property in one of the golden miles
in the world by a Chinese enterprise based in Hong Kong, which is
symbolic for the business prosperity of Lee Kum Kee. Secondly, the
timing of this transaction was remarkable. It was at a period when the
value of the UK pound was on the decline and the British real-estate
market was in the doldrums following the Brexit referendum. Indeed,
the long-term reappearance of a premium property may always be
attractive; nevertheless, it was a bold and visionary decision to invest
heavily during the downturn of the market. The Chairman of the Lee
Kum Kee Group is Lee Man Tat, a nearly 90-year-old entrepreneur,
who has experienced countless adversities and predicaments in life yet
remains highly energetic and aggressive in business pursuits.

Although Lee Man Tat is a man well-known to the industries of
condiment and health products, nevertheless, unlike the household
name of Lee Kum Kee, he is a less familiar figure to the general public.
In fact, Lee Man Tat is the soul of Lee Kum Kee, who inherits and
continues the heritage of Lee Kum Kee; it is also he who has steered
the company from small to large, from Macau and Hong Kong, to the
overseas markets, and finally, to the whole world and outer space — its
products accompanied the astronauts of Shenzhou IX, X, XI and the
International Space Station to space as part of their food condiments.

The ancient Greek philosopher Archimedes once said: "Give me a
fulcrum, and I will be able to move the earth." Lee Man Tat understood
where the fulcrum of his career was, and, based on this, has achieved
a luminous life. He not only laid a solid foundation for the long-term
development of Lee Kum Kee, but has also successfully forged a

world brand. Lee Kum Kee under his leadership has advanced from merely producing oyster sauce and shrimp paste to a wide variety of condiments today, thereby enriching the daily diets of millions of households. Moreover, Infinitus — an extended brand of Lee Kum Kee — is committed to offering a multifarious range of health products to promote Chinese medicine and health regimen, which has matched the new trend of pursuing a healthy lifestyle amongst consumers across different strata.

In short, Lee Man Tat has made himself a legend through the fulcrum of his career. His extraordinary life story, however, has not been known, as yet, to society at large. Lee Man Tat was initially named "Lee Kin Tai" by his parents. Such a special name was given to him on two grounds: the joy and hope that his birth had brought to his parents; and the wish that he could grow up healthily. Lee Kin Tai was a frail and sickly baby, thus, his distressed parents worried about his viability very much. They blessed him with the name of "Kin Tai", which literally meant "watch (him) grow up big". However, intriguingly, another literal meaning of "Kin Tai", which could be interpreted as either "seeing something big" or "seeing the big picture", in fact, reflected a unique ability of this baby in the future — an ability to see the big picture from the small detail. This was beyond the expectation of his parents.

Notably, two personal attributes of Lee Man Tat could be summarised from his legendary life. They are, first and foremost, his vision to "see the big from the small"; and secondly, his ability to grasp opportunities in crises. Implications of these two attributes or personal styles of Lee Man Tat could be traced in three significant and interlocking aspects of his life, namely his upbringing, his business career, and his personal culture.

Life is intriguing and full of surprises. Lee Man Tat, as a sickly and feeble infant did worry his parents, yet as it happens, he has shown a great tenacity for life. His parents were highly distraught by the poor health of their "first child" and tried their best to seek medical treatment and prayed to the almighty God. Lee Man Tat's health gradually improved under their tender care and he grew up as strong as an ox. He got married later and expanded his family, and is now surrounded by a large number of children and grandchildren. We could hardly imagine that Lee Man Tat, a spirited octogenarian, was once a child that deeply worried his parents concerning his viability. Hence, the literal meaning of Kin Tai — "watch (him) grow up big" — had been realised, and it was, indeed, a story of turning crisis into opportunity during his childhood.

Lee Kum Kee's products can now be found all over the world and its investment activities cover a vast geographical spread. However, its path of development from small to big was by no means a smooth one as some might imagine today as it was full of setbacks. In retrospect, when Lee Man Tat was first involved in Lee Kum Kee's business, the company was no more than a diminutive oyster sauce store. Lee Kum Kee's foundation was laid by his grandfather and father, but it remained a small firm with the layout of its store as a "shop at the front and a workshop at the rear". The young Lee Man Tat joined Lee Kum Kee regardless of its size, and he started out in junior positions, paying attention to every detail of Lee Kum Kee's operation, including its procurement of raw materials, its process of production, its quality assurance, its sales and marketing, as well as its relationship with clients and consumers. He fully understood how minute details could affect major outcomes, and how the big picture could be envisaged from clues provided by the small details. He also believed that so long as we worked harder, petite businesses could also thrive if given the right environment. These "ordinary" attitudes of life and the basic principle in dealing with people and matters have, in fact, shaped his unique philosophy of "100 - 1 = 0", as well as motivated him to lead the company from small to large that finally morphed into a multinational group.

Understandably, this is a course full of twists and turns as well as setbacks. Difficulties may emerge from challenges in running and managing both Lee Kum Kee and the family, or from ecological changes of the industry and society, as well as politico-economic changes in the region and throughout the world. Lee Man Tat consistently stuck to his conviction in each and every case of distress, never being intimidated by any adversity or crisis. He tackled challenges calmly, seized opportunities and held them tight through critical moments; meanwhile, he worked even harder. As a result, every time when the family and the business encountered a serious problem, it was Lee Man Tat who composedly turned that crisis into an opportunity with his amazing insightful and daring decision, thus ensuring an incessant breakthrough for both the family and the business.

Oyster sauce is, to some, a mere condiment that could not satisfy hunger, and is, thus, a rather insignificant and dispensable commodity in their eyes. They could hardly associate oyster sauce with the Chinese culture, and in particular, the Chinese culinary culture, Chinese medicine, and health regimen, which are accumulated wisdoms of over several thousand years. In fact, oyster sauce could project the uniqueness of the Chinese culinary culture and reflect the rich heritage of the Chinese health regimen. This "insignificant" oyster sauce could substantially enrich the flavour of our meals, and is, hence, indispensable to millions of households.

Precisely because Lee Man Tat could see the big from the small and realise the immense potential behind such a tiny drop of oyster sauce, he always took the business seriously and warily as a mission to inherit and promote Chinese culinary culture. Following the footprints of his ancestors, Lee Man Tat first entrenched the company's sales network in Macau and Hong Kong, and later expanded such network to the

North and South Americas, Australia, and Southeast Asia to cater to the increasing demands from the overseas Chinese communities. The move subsequently popularised Lee Kum Kee's oyster sauce in many places worldwide, and along the way, also contributed to the fame of Chinese cuisines and culinary culture. He was likewise hit by crises and setbacks during the course, but his ability to turn crises into opportunities as well as his perseverance eventually prevailed.

Compared to oyster sauce, the great variety of Lee Kum Kee's health products apparently reflect the more eye-catching fields of Chinese medicine and health regimen, and as we all know, they have a long history. Their philosophy of accommodating the rules of nature and fortifying the fundamentals of the human body is, in particular, well-regarded by many. Based on his wealth of experience in expanding the oyster sauce market, Lee Man Tat had become even more insightful when he established Infinitus, and was therefore able to grasp opportunities in the process with a more measured approach and accurate judgments. As a result, Infinitus has excelled and stood out in the Lee Kum Kee Group. Indeed, Infinitus did not enjoy a smooth course of development either, yet Lee Man Tat, as the common asset of the two businesses of Lee Kum Kee, could always handle any trouble with his ability to grasp at opportunities.

Lee Man Tat has devoted his entire life to promoting sauce and health products with perseverance. Apart from making these Lee Kum Kee products ubiquitous worldwide, he has also contributed to the popularity of Chinese culinary culture, medicine, and health regimen, which in fact, reveals his enormous confidence towards the Chinese culture. Lee Man Tat, in his adolescence, once experienced the critical danger of China and the Chinese people under the Japanese invasion and occupation. The declining and feeble China at that time had also prompted many countrymen to turn their backs on their own culture and passionately espoused Western culture. However, Lee Man Tat remains proud of the Chinese culture regardless of whatever flaws there might be. He sets an example in his own family and business, aggressively commits to their development and modernisation, and throughout the course demonstrates to others that Chinese culture could indeed incorporate modern elements and subsequently move towards a resilient, inclusive, and vigorous modernisation. In short, there is a role to play for the Chinese culture in the national rejuvenation.

Lee Man Tat achieved something great because he was able to see occurrences at different levels and to consider others' interests. That implies a meticulous attention to details, a willingness to bear immediate hardships, and a lasting appreciation of the kindness from others. Moreover, he is talented in grasping opportunities, steadfast in his conviction, highly confident, and not beset by resentment. He is also benevolent, willing to give back to the community and contribute to his country. These unique qualities and styles of Lee Man Tat are not something others could easily learn, copy, and then practise, and thus, are imperative for a written record to enlighten the generations that follow. Younger generations of his family and the general public may then have an insight into the hardships of his entrepreneurial process, as well as his personal attributes and styles.

It is, of course, not easy to comprehensively present Lee Man Tat's colourful and legendary life, particularly in outlining his perceptual outlook succinctly, let alone to also incorporate those turbulent historical changes of the times into our narrative. In the face of such challenges, we encountered difficulties, obstacles, as well as criticisms. We were fortunate enough to finally complete this research project with the help of many friends and institutions. We must therefore express our deep gratitude to them for their support and assistance.

First and foremost, we would like to extend our warmest thanks to

Mr. Lee Man Tat and Mrs. Lee Choi May Ling. They not only shared their life experiences and views of things with us in each interview and personal communication, but also their concerns towards social and national developments. Such interaction and exchange of thoughts are truly memorable.

In the same way, we would like to thank other members of the Lee family — Eddy Lee, Elizabeth Lee, David Lee, Charlie Lee, Sammy Lee — as well as other interviewees, including Bernard Yiu, Meng Suhe, Wei Xiangyun, Chen Ying Ming, Feng Enyuan, Yang Jiaqing, Tony Mok, Zhao Shuangwei, Li Quansheng, He Xiansong, Li Jiuru, Li Baihan, Li Guoli, Fan Xiaogang, Li Zhinong, Zhen Wonan, Yang Jieming, Zhong Anxiang, Ho Shiu Chung, Chang Chin Vai, Tsoi Cheong Ming, Mr. & Mrs. Cliff Ho, Raymond Lai Ming Kee, Vincent Ko Hon Chiu, Harry Yeung Kwok Chun, Lam Yu, He Yongxiang, Stephen Chung, Michael Sze Cho Cheung, Tang Fook Chuen, Choi Cheong Too, Samuel Yu Kwun Chark, Chiu Sin Ching, Chow Sing Lam, Ho O Sang, Hui Chi Hing, Pao Pui Li, and Mr. & Mrs. Loi Kam Sheung. We really appreciate their patient sharing, sincere cooperation, and their selfless dedication very much.

We would also like to thank the Special Collections of The University of Hong Kong Libraries, Government Records Service of the Government of HKSAR, the Lee Man Tat Secondary School in Qibao, Xinhui, and the Lee Shiu Nan Memorial Primary School in Nanshui, Zhuhai. These institutions have greatly facilitated our archive research and information gathering at various levels.

Finally, we would like to express our heartfelt thanks to all the team members from Lee Kum Kee, namely Mr. Tang Fook Chuen, Mr. Choi Cheong Too, Mr. Cletus Lau, Mr. Dodie Hung, Mr. Bridge Mak and Mr. Wong Ping Yuk. They had made a great effort and offered kind advice to this study. Without their support, the content of this book may be less substantial and incomplete; and without their toil and sweat, we may even have failed to accomplish this project. We would also like to thank Professor Lee Chack Fan for his kind assistance, as well as Professor Jao Tsung-I for his calligraphy inked as the book's title. Sadly, Professor Jao passed away recently. And last but not least, we thank Ms. Anita Lai, Corporate Affairs Manager, and Ms. Kelly Lam, Executive Assistant of Lee Kum Kee International Holdings, for their highly efficient administrative support.

Despite all the great help from friends and institutions, mistakes and omissions in the book are inevitable, as we were unable to fully grasp the dramatic political transformation throughout this long and tortuous period of history, not to mention the sophisticated process of an enterprise's inheritance, as well as the ups and downs of one's life. Although we did strive to seek evidence and confirmation, some undecided and ambiguous points in the book remain unresolved. We therefore sincerely appeal for feedbacks and criticisms from readers to enrich and substantiate our future research.

CONTENTS

CHAPTER 7
INVEST IN CHINA AMIDST CRISES

CHAPTER 8
GO WORLDWIDE IN CRISES

CHAPTER 9
PREEMPTING FAMILY CRISIS

CHAPTER 10
WISDOM OF A TIDE CHANGER

Thank my grandparents, my parents, my wife, my children, and grandchildren, my business partners, friends and colleagues. They are the ones that inspire me and enrich my life.

Lee Man Tat

The loving couple of Lee Man Tat and Choi May Ling: one year after they had tied the knot.

The couple and their five children: crossing Victoria Harbour by ferry.

Chang Chin Vai (1st right) has followed the visionary Lee Man Tat (2nd left) throughout the latter's career and assisted in the management of the Macau Lee Kum Kee Store in later years.

A subtle contrast of cultures: group photo of the traditionally dressed Lee Shiu Nan with Lee Man Tat and his friends, who all wore Western formal suits.

Sweet smiles when travelling.

Became grandparents: amused by
their warm-hearted grandsons.

Gracefully posed in front of the Sha Tin Floating Restaurant.

A "Wild" trip to Northern China: mounting a camel against the backdrop of the Great Wall.

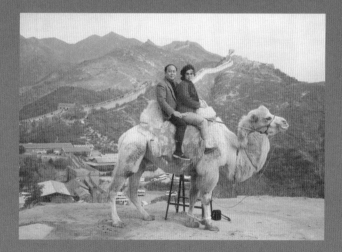

Inspecting the display of Lee Kum Kee's products on supermarket shelves to observe their market penetration.

To Lee Man Tat, business partners are always beyond business. They are his good friends.

Like father, like son: Lee Man Tat went to Europe to promote Lee Kum Kee in the 1980s with his sons Eddy Lee (1st right) and Charlie Lee (1st left). The father and sons were alike in their clothes as well as their emotional expressions.

Procurement of fresh shrimps on shore is an important step to ensure the quality of Lee Kum Kee's shrimp paste. Lee Man Tat is with his daughter, Elizabeth Mok, at a pier to inspect the procurement of fresh shrimps.

The warm relationship between Lee Man Tat and the former Chief Executive Tung Chee Hwa was apparent as they revealed in their gestures.

Received the "Southern-Top Ten Chinese Philanthropists" Award for his longstanding contributions to charity and community service.

A deserving winner of the "Industrialist of the Year" Award.

Attended the "Third China Fair for International Investment and Trade" and posed with Wu Yi, the then Minister of Foreign Trade and Economic Co-operation of China.

Received the Coutts Prize for Family Business as a recognition of their superb family governance.

Expression of appreciation to astronauts for taking Lee Kum Kee's products with them into outer space.

Contributed to the building of Lee Shiu Nan Memorial Primary School to commemorate his father. The couple always enjoyed their moments with students upon their graduation.

Start Off Ceremony of the "2006 Lifeline Express Car Expedition to Sichuan". The couple thoroughly enjoyed participating in charity work and public service.

The Lee Kum Kee management photographed at the Tai Po Headquarters in 1995, right after the unveiling ceremony of the white stone statue of Lee Shiu Nan.

The blissful banquet for the 60th wedding anniversary cum 85th birthday of the couple.

CHAPTER

1

Oyster farming and harvesting in the early days. Farmers had to open the rugged and tightly-closed oyster shells to collect the delicious and tender oyster molluscs inside.

PRELUDE TO A LEGEND

Fleeing Qibao from the furor of local gangsters, Grandpa finally settled in Nanshui. He set up a little teahouse there and struggled to make ends meet. That teahouse served dried oysters as well. One day, Grandpa accidentally overcooked oysters to a brownish thick paste in the wok. Hesitated in pouring it away, he tasted the paste, and found it surprisingly delectable. He then sold it in his teahouse, and named it "oyster sauce", hence, the historical debut of oyster sauce and, the brand, Lee Kum Kee...

Lee Man Tat

INTRODUCTION

It was a sunny day in October 1990. Lee Man Tat and Choi May Ling, together with a small group of their friends and relatives, headed to Nanshui, Zhuhai from Hong Kong for a meaningful occasion — the inauguration ceremony of Lee Shiu Nan Memorial School. Zhao Guocang, Party Secretary of the Qibao Town, Xinhui County also attended that occasion, with an intention to persuade the couple to visit Qibao immediately after that event.

Nanshui and Qibao, both as parts of the Pearl River Delta, are only some 80 kilometers apart. Today, it takes just one and a half hours drive. Yet, back in October 1990, even though the Hong Kong guests were escorted in Mercedes-Benz by the Qibao authority, the journey was deemed unexpectedly tortuous. Qibao was indeed isolated from the outside world by its surrounding watercourses, without any bridges connecting. Travellers from Nanshui to Qibao, for instance, had to make a total of seven river crossings via rafts. By the time they finally made it to Qibao, it was a starry night to greet them, followed with mental and physical fatigue.

Along the pier, local primary school students had been waiting long with empty stomachs and bouquets, up to the town office. Li Guoli, Deputy Town Mayor, taking charge of the welcoming event, signalled students to cheer, and demonstrated a grand lion-dance performance amidst deafening firecrackers. However, Lee Man Tat was soon aware that these kids were hungry and equally exhausted for waiting so long. Frowned upon the scene, he requested Li Guoli to let the kids go home immediately, saying that if they summoned school kids for such a line-up again, he would not return for a second visit.

Apart from the grand welcoming, Lee Man Tat and his entourage were also surprised by the signboard of Lee Kum Kee hung around the pier, as well as the painting of the company's famous advertising slogan — "Lee Kum Kee, your indispensable sauce at every meal" — on walls along their way. Given the fact that back in 1990, Lee Kum Kee's business was insignificant in Mainland China, it was absolutely hard to believe that these signs of Lee Kum Kee's "popularity" in Qibao were solidly based on its retail strength. Obviously, these were "art works" crafted by the local cadres. When the group finally settled down in the town office for a dinner, they could still smell the firecrackers, and the school kids remained standing. Lee Man Tat furiously turned to the local cadres: "Hey, what's wrong with you guys?" Only after these words that the children were finally dismissed.

Nearly three decades later, most of these local cadres, while accepting face-to-face interviews with us, admitted that all these were laboriously set and planned to lure Lee Man Tat. They had learnt about his intention to invest in Nanshui. In comparison, Qibao was a much less attractive choice for the investors, and they knew it, but still, they dreamed for a solid investment from Lee Kum Kee. They were ready to generously offer a chunk of land for the company, plus a promise of comprehensive coordination. Above all, they also believed that they held a trump card in persuading Lee Man Tat, namely his "hometown nostalgia". In fact, Lee Man Tat's somewhat surreal overnight trip to Qibao in 1990 was his very first to his "hometown". Born and raised in Macau, Lee Man Tat is indeed a son of Macau, but Nanshui and Qibao are two equally momentous places to him and to Lee Kum Kee: Nanshui is where Lee Kum Kee's founder, Lee Kum Sheung, came up with his very first wok of oyster sauce, which was also the first in China, and hence Nanshui marks the starting point of Lee Kum Kee; whereas Qibao is where Lee Kum Kee hailed from, and therefore an ancestral hometown for the Lee family.

Back to the starry night in Qibao in October 1990, the little town soon returned to tranquility after the grand welcome and dinner event, with dogs barked occasionally. Lee Man Tat, the then 61-year-old, couldn't fall asleep quickly in a hotel nearby. He was deeply touched by the locals, or in a sense by his fellow countrymen, but when it came to the idea of investment in Qibao, the "seven rounds of river crossing via rafts" was a formidable traffic barrier to any rational and pragmatic businessman. He consulted Tang Fook Chuen and Choi Cheong Too, two of the senior executives of Lee Kum Kee who travelled with him to Qibao, and received advice similar to his viewpoint. Choi Cheong Too, for instance, bluntly commented that in terms of timing, geographical advantage and human relations, "Qibao is a place excellent in human relations, yet geographically disadvantaged." Hometown nostalgia eventually failed to beat the reality of a poor infrastructure, and Qibao was unable to turn Lee Man Tat around in 1990.

The following few years saw Lee Kum Kee continued to discuss with the authority of Zhuhai City regarding its investment proposal in Nanshui. Foshan, Shenzhen, Guangzhou's Conghua and other cities/ districts in the Pearl River Delta were also eager to solicit Lee Kum Kee. These locations had all been considered and inspected personally by Lee Man Tat and his eldest son, Eddy Lee. However, deep in his heart, it seemed that Lee Man Tat remained hesitated between the two options of Nanshui and Qibao. By December 1992, Qibao began its substantial transformation thanks to the official launch of Wong Haking Bridge, which connected Qibao to the outer world and hence partially alleviated its transportation plight. The bridge was donated by another Hong Kong industrial entrepreneur, Haking Wong, who was also tracing his ancestral home to Xinhui as well. Meanwhile, Zhuhai's Party Secretary cum Mayor Liang Guangda wasn't too keen on Lee Kum Kee's investment proposal in Nanshui, as he was pre-occupied by plenty of others on his desk at the time.

Lee Man Tat returned to Qibao together with his family in 1994, with a specific purpose to pay homage to his ancestors. During the stay, they also took a closer look at the ancestral home of Lee Kum Sheung in Chongli Village and the vicinity. Amongst the available locations for his investment plan, Lee Man Tat finally decided to set up a mammoth production base of Lee Kum Kee in Qibao, even though Qibao remained a relatively inferior choice. The company then swiftly commenced building its factory infrastructure in Qibao in March 1995, and completed the first phase of the construction plan by April 1996. Since then, the production base has delivered various products of Lee Kum Kee.

"I selected Qibao with a wish to keep ties between Qibao and our family. I want all of you to be able to come back here regularly..." Lee Man Tat once explained his choice of Qibao to Eddy Lee in such way. Indeed, he has been very serious about that. Since the "Shiu Nan Lodge" situated within the production base was ready, Lee Man Tat has allocated all those spacious rooms on the Lodge's upper floor to each of his children, so as to facilitate their visits to Qibao. Today, Qibao has been the largest production base of Lee Kum Kee, in particular the making of soy sauce and its associated products. The base covers an astonishing 2000 mu (approximately 1,333,334 m²) (1)of land, with an impressive display of giant tanks purposely designed for soy fermentation, thousands in number. These are cylinders that allow natural fermentation of soybeans under the sunlight of Xinhui. Albeit time-consuming, such a traditional South-China way of fermentation yields better soy sauce.

Lee Kum Kee is today the largest enterprise in Qibao, as well as the prime employer and tax payer. Both Lee Kum Kee and the Lee family have left an imprint deeply upon this little town. Apart from the production base, Lee Man Tat has donated public facilities built around his ancestral home in Chongli Village. These include the Lee Kum Sheung Memorial Pavilion (2000), Lee Man Tat Square (2005) and Lee Shiu Nan Memorial Pavilion (2006). He has also been dubbed a member of the "seven sages" of the Lee clan of Qibao, and his virtuous deeds are highlighted together with his portrait on a wall of the Lee Ancestral Hall. Moreover, throughout these years, the Lee Kum Kee Group has funded several major educational and infrastructure projects in Qibao which, amongst others, are Lee Man Tat Secondary School (1998), Lee Man Tat Bridge (2014) and Infinitus Bridge (2017). These projects have in no doubt contributed significantly to the locals' wellbeing.

Fatal Threat to Lee Kum Sheung

A prelude to Lee Man Tat's legend, therefore, should begin in Qibao, and all the way through Nanshui, Macau, Hong Kong, Mainland China and the world. These tortuous footprints not only reveal an unusual story of a family business developed from just a little workshop to the present giant, but also reflect the dramatic twists and turns of the modern Chinese history. The story indicates — as in many others — that life is full of ups and downs, but in the face of great difficulties, a sunny day may follow if the family could finally overcome them. Otherwise, it may be doomed to irrelevance.

Lee Man Tat's grandfather, Lee Kum Sheung, was born in 1868 during the late Qing Dynasty of China. He grew up in Chongli Village, a place by then under the Qibao Township of Xinhui County, Guangdong Province. Lee Kum Sheung's father, a local farmer, died early, thus he was raised by his widowed mother. Today, we know little about the life of Lee Kum Sheung, but according to Lee Man Tat, Lee Kum Sheung learnt Chinese martial arts since his childhood, and as a skilled martial arts practitioner, he always stood up against injustice.

A brief biography of Lee Kum Sheung published in the *Zhuhai Qiaowuzhi (Overseas Chinese Affairs of Zhuhai)* (2009) notes that Lee Kum Sheung, at the age of 18, attended and excelled in the official examination for military personnel held in Huicheng, Xinhui. He then returned to Chongli Village and waited for official recruitment. However, his outstanding performance in the examination had drawn jealousy from a rich young man of the same village, who, then "stirred up troubles, called upon friends to besiege and beat Lee Kum Sheung." Lee Kum Sheung couldn't help but was forced to flee Qibao. He headed for Nanshui Market of the Xiangshan County, which is today's Nanshui Town of the Zhuhai City. After settling down in Nanshui, he surreptitiously returned to Chongli Village in the following year to fetch his mother to Nanshui.

Another version of the story, which is widely told amongst the locals as well as the family members, is that young Lee Kum Sheung had offended the local bully out of righteous indignation. He was then informed by a close friend that the bully would attack him in a certain late night, and he barely escaped the fatal strike with his mother. It was imaginably a very hasty move, but lives saved.

Discrepancies found between these two versions of story are trivial. Given the backdrop of weakening authority of both the central and the local governments during the late Qing Dynasty, an underprivileged young man and his widowed mother, confronted with such a fatal threat, could only run away. Lee Kum Sheung's martial arts skills obviously might not be enough to assure his survival in the strike. His swift and rational decision to flee his hometown and search for other 'way out has rewritten destiny of the Lee family.

It was said that after they had left Chongli Village, Qibao, the mother and son took a boat along the Tanjiang River and sailed southward, eventually settled in the southern tip of the Pearl River estuary, a seashore town called Nanshui which was anciently known as Langbaiao. They fled hurriedly and so brought almost nothing with them. Once settled, Lee Kum Sheung had to work diligently for a living. He was once a porter, a woodman, and an employee in oyster farming, paddy farming and fishing (Zhang, 2009). As an expansive and upright young man, Lee Kum Sheung mingled well with the locals. He even became a sworn brother of Lee Man Chan from Qibao who had always cared for him in Nanshui.

A year or two later, the industrious and thrifty Lee Kum Sheung had saved some money and decided to open a teahouse on an open ground just next to his home. Lee Man Chan was supportive enough to join

Lee Kum Kee's founder Lee Kum Sheung and his wife Leung Tong.
Their traditional attire and melancholic looks might well reflect the
historical background and conditions of the earliest development of
Lee Kum Kee.

hands in setting up and running this petit business. The little teahouse was located in the east of Nanshui Village, which is today the western side of the Lee Shiu Nan Memorial Primary School. Albeit humble, the shop was where Lee Kum Sheung embarked on his entrepreneurial adventure, and it laid the foundation for his oyster sauce business later.

"An Affair with Oysters" for Five Generations

Oysters mean a great deal to Lee Man Tat and his family, to the extent that their business and fate can be described as "an affair with oysters" for over five generations, with the enchantment sustained. Lee Man Tat's extraordinary career is particularly linked to this ordinary aquatic creature. Given the significance of oyster enchantment in our legend, it deserves a brief introduction to its features.

Oyster is the common name for certain species of bivalve molluscs that live in marine or brackish habitats. Being a seafood widely and popularly consumed by the Chinese, oysters have been called different names in different Chinese dialects and regions. Oysters feed on plankton, and brackish waters around the Pearl River estuary are especially suitable for plankton breeding. It is therefore not surprising to find several notable oyster-cultivating centres along both the east and the west banks of the region, for instances, Shenzhen's Shajing and Hong Kong's Lau Fau Shan on the east, and Macau and Zhuhai's Nanshui on the west. Historically, locals of the region not only cultivate and consume oysters, but are good at adopting oyster shells in their house constructions. It is also noteworthy that the two ancient names of Macau, Haojingao and Haojiang, indicate the significance of oysters in the local lives of Pearl River estuary, as the Chinese character of "hao" is in fact derived from the character of oysters.

Archaeological discoveries and historical documentation both show that the Chinese might have learnt the skills to cultivate oysters as early as the Han Dynasty, which was around two thousand years ago. Amongst others, "chazhu yangli" — cultivate oysters by inserting bamboo cane into the shallow sea bed — was an early cultivation method mentioned in ancient Chinese literature. Such a ground-based approach (vis-à-vis the suspending approach) has been widely adopted and advanced in the Pearl River estuary ever since the Song and Yuan Dynasties. The intriguing oyster farming in the New Territories, Hong Kong had also prompted Yeh Ling-feng, a renowned writer on Hong Kong's folklore, history and geography, to write a piece of exquisite account describing its farming skills in the 1950s.(2)

Oyster molluscs are delicious, highly nutritious and easy to digest, hence there are myriad of oyster lovers on all sides of the world, the Chinese and the Westerners alike. Rich in protein, oysters have even been dubbed "sea milk" by the Europeans.(3) However, the Chinese, unlike their Western counterparts, tend not to consume raw oysters, as oysters are regarded as something "cold" in nature according to theory of the traditional Chinese medicine. They will either consume oysters after cooking, or turn it into dried oysters (haochi) after processing. Different ways of processing will lead to different types of dried oysters: one dried up under the burning sun without boiling, while another after boiling. Either way, it takes at least several sunny days to have them done.

There is of course one more exquisite way to consume oysters, which is to keep boiling them until the soup becomes highly concentrated, then to add some salt to make it as oyster sauce. The sauce is undoubtedly a popular condiment amongst the Chinese today, however, if we search through historical documentation written before 1888, here we are referring particularly to local chronicles edited by county or township authorities in the vicinity of the Pearl River estuary, it seems that no earlier note on oyster sauce could ever be found. Local chronicles occasionally did mention oysters, for instance, *Xinanxian Zhi (Chronicle of Xinan County)* once noted that "the sea of Helan and the beach of Baihe both produce oysters. It is the locals that divide the land and cultivate them, in their so-called oyster farms…", this paragraph nevertheless related only to oyster farming.

Inventing Oyster Sauce

The Lee family's affair with oyster could be traced back to 130 years ago, i.e. 1888, at the little teahouse run by Lee Kum Sheung in Nanshui. It is said that oyster sauce was accidentally invented, just like the discovery of gravitational force by Isaac Newton, was inspired by a falling apple, or the discovery of Penicillin by Alexander Fleming due to an unexpected blob of mold growing on one of his petri dishes. These discoverers kept exploring implications of such discoveries, until they finally came out with something marvelous.

Oysters abound in Nanshui. Lee Kum Sheung's little teahouse, therefore, also served oysters, but what he originally offered was dried oysters. One day, while boiling oysters for the making of dried oysters, a busy Lee Kum Sheung inadvertently left his wok on the stove much longer than needed, until the soup within turned from a milky white fluid into a brownish black paste, which looked unpleasant but smelled alluringly.

The waste of overcooked oysters dismayed Lee Kum Sheung, but the tantalising aroma from this brownish paste prompted him to scratch a bit from his wok and tasted it. It was unforgettably yummy. Lee Kum Sheung was not sure if he had biased taste buds, so he asked his mother and friends in the village to also taste the paste, and they unanimously confirmed its delicacy. Thrilled with such discovery, the young entrepreneur began to conceive ways to commercialise this brownish paste of oyster.

After a series of trial and error, Lee Kum Sheung eventually developed a better approach to production of the sauce. He found out that a 50kg of oyster soup could be boiled down to around 5kg of essence, which should then be seasoned with salt and filtered to remove impurities precipitated on the bottom of wok. He ingeniously named it "oyster sauce" with the brand of "Lee Kum Kee" ("Kee" as a suffix here refers to "brand"), and started to make it a product for sale. His grandson, Lee Man Tat, depicted the debut of oyster sauce over a century later in this way:

> *Fleeing Qibao from the furor of local gangsters, Grandpa finally settled in Nanshui. He set up a little teahouse there and struggled to make ends meet. That teahouse served dried oysters as well. One day, Grandpa accidentally overcooked oysters to a brownish thick paste in the wok. Hesitated to pour it away, he tasted the paste, and found it surprisingly delectable. He then sold it in his teahouse, and named it "oyster sauce", hence, the historical debut of oyster sauce and, the brand, Lee Kum Kee...*

Since then, Lee Kum Sheung collected oysters from various sources in daytime, and processed the sauce at home in the evening. He then sold it on the street. This branded savory sauce began to bring him steady income, and gradually established itself as a distinct taste amongst the locals. The business finally enabled Lee Kum Sheung to get married, fulfilling his mother's long-awaited wish. His wife Leung Tong came from a wealthy family in Xiangshan (today's Zhongshan). The marriage thus shouldered him with heavier responsibility, as well as a stronger motivation to succeed.

The popularity of Lee Kum Kee oyster sauce soon begot imitation. Oyster sauce, after all, was a product that could come out from any family workshop, so long as they knew the way to boil oysters, but the quality varied. Oyster sauce workshops popped up in Nanshui, and the local oyster sauce industry allegedly reached its height by having more than ten shops in number. The sauce became widely known in the Pearl River Delta, so much so that when merchants from Jiangmen, Shiqi and Guangzhou came to Nanshui for its fishery products, they often purchased oyster sauce as well. A regional market of oyster sauce started to take shape, and Macau notably became the major importer of oyster sauce from Nanshui.(4)

Given the chaotic and unhealthy competition, how did Lee Kum Kee excel others? Lee Kum Kee's early day performance was made unavailable to be traced due to the 1902 conflagration in Nanshui, which had all the records destroyed. However, fragmental information collected from different sources and our interviews with senior members of the Lee family could still help us to figure out its early trajectory.

At the very beginning, Lee Kum Kee was nothing more than a little workshop with oyster sauce as its only product. But it seemed to enjoy a reputation of "the best quality" since its debut. Such a fame did not come out of thin air. Stories and lessons pass from generation to generation within the family include principles of production strictly abide by Lee Kum Sheung: he would, first of all, apply only fresh oyster molluscs and high-quality ingredients in making the sauce; and secondly, he would always ensure a hygienic environment throughout the process of production.

The chivalrous character of Lee Kum Sheung might have also facilitated his career. Over the years, he remained an upright person with compassion. His benevolent act of writing off debts of certain customers of his teahouse is still widely told and praised by the locals today. On the eve of every Lunar New Year, Lee Kum Sheung would simply take all the debit notes out and burn them in the stove. Customers of his Nanshui teahouse were mainly fishermen, and he allowed some destitute customers to eat and drink on credit. The

Chinese customarily repay their debts before the Lunar New Year, and if they don't, it usually means they really couldn't make it, or are currently facing difficulties due to unexpected events. For instance, in Lee Kum Sheung's case, a fisherman who never returned to his family from the sea. Either way, Lee Kum Sheung would just burn their debit notes off. Lee Man Tat's father, Lee Shiu Nan, had repeatedly inculcated this story in his mind since his childhood, and always with a quote directly from Lee Kum Sheung: "We shouldn't burden the next generation with their father's debt." Lee Kum Sheung embedded values of "quality first" and "integrity" into the genes of Lee Kum Kee, which had very much contributed to its early days' success.

Finding a Way Out After Fire

Although far from affluent, Lee Kum Sheung's happy marriage and his booming business had made him a subject of envy in Nanshui. The couple began to have children, and the kids warmed up their home.(5) In less than two decades after he fled Qibao, Lee Kum Sheung had effectively turned himself from a casual labour to a teahouse co-owner, and later to the founder of an oyster sauce business who ran a shop in the busiest Nanshui Street.

Lee Kum Sheung's mother, however, passed away at the height of his career in Nanshui. It was particularly mournful to Lee Kum Sheung, for he was more able than ever to observe filial piety.

Lee Kum Sheung suffered another blow in 1902 when everything of his life was just on the right track. It was a devastating disaster out of the blue, and had Lee Kum Sheung unprepared to make a difficult and critical decision. Both his family and Lee Kum Kee's trajectories of development changed accordingly.

It was a day in 1902, not long after Leung Tong gave birth to the couple's third son, Lee Shiu Nan (i.e. Lee Man Tat's father), Nanshui Street was ravaged by a sudden conflagration. The fire engulfed street shops one by one, including Lee Kum Kee's. Lee Kum Sheung, totally impotent in front of the shocking blaze, could only wait for the flaming of almost all of his properties amassed over the last decade to ashes.

It was very much like a rolling back to ground zero to Lee Kum Sheung. He was once again a have-not, but where should he head for this time? The decision to flee to Nanshui was easier for a young Lee Kum Sheung to make as he was then a bachelor, only needed to take care of his widowed mother. But at the moment, he was older, married, and a father of four. Should he stay in Nanshui for a reboot, or move to Macau for a greener pasture?

After much deliberation and a full discussion with his wife Leung Tong, Lee Kum Sheung finally decided to head for Macau, along with his family. The decision implied a big change as they were going to reset their lives in a different and unfamiliar neighbourhood, a place where the East met the West, and where the locals were purportedly materialistic and less benevolent.

Macau was a port city opened up to the outer world much earlier than Nanshui. It had long been a rendezvous for the Southern Chinese and European traders, as well as a melting pot for both cultures. Lee Kum Kee, albeit based in Nanshui, had exported a significant quantity of its oyster sauce to Macau, and from Macau reached out to a few overseas Chinese communities. A foothold in Macau might therefore imply a larger market for Lee Kum Kee.

Lee Kum Sheung visited his mother's tomb with his family before setting foot on another journey. Looking back, his decision to leave Nanshui was rather rational: on business opportunities, Macau had become the major destination for Lee Kum Kee's oyster sauce; on

Early days of Nanshui: a tranquil fishing village crowded with houses.

geographical accessibility, Macau was one of the few big cities that could be reached by passenger steamers directly from Nanshui. A deeper strategic implication of the move upon Lee Kum Kee, however, could become clearer only a century later.

Conclusion

"Plum blossoms come from the bitter cold." Lee Man Tat contributes immensely to Qibao over the years in terms of local employment, local constructions and tax revenues, which have made him a popular and highly respected figure in Qibao. These contributions are possible due to his excellent leadership of Lee Kum Kee. Lee Man Tat has greatly expanded the foundation of this century-old family enterprise founded by his grandfather and strengthened by the second generation via a series of major breakthroughs. It is now ready to be passed on to the fourth and the fifth generations. Lee Man Tat is indisputably a pivotal figure of the family heritage, and has facilitated the extension of the family's affair with oysters.

Lee Kum Sheung passed away shortly before Lee Man Tat was born. The grandfather and the grandson, though never had a chance to meet each other, share an amazing trait of personality: both of them, in the face of immense crisis, could always see through hidden opportunities from crisis, and have the situation turned around. This is the prelude to a legend.

1 A mu equals to approximately 666.667 m², or 0.164 acre.

2 "Most oyster farms in the New Territories are located in either the seaside or the estuary, where fresh water of the rivers meets salty water of the sea. These are the most ideal locations for oyster farming. The sea bottom concerned should be sandy, with a mix of sludge. Oysters will not easily get "fat" without these sludge, but if the sludge layer is too thick, it will adversely affect their breeding as well. Oysters could not stand for strong wind, and dislike sunlight, therefore the oyster farm is best set in a position that sheltered from the wind. Powerful typhoons that stir up big waves are the greatest threat to oyster farms, but if the water is too shallow and so oysters are let to expose directly under the sun, it causes trouble as well. Oyster farmers in the New Territories usually drop bricks, tiles, pottery pieces and empty oyster shells into the sea bottom of the farm, in order to create the best "homes" for oysters. They will burn the bricks red before dropping, claiming that it will facilitate oysters' breeding. I think it works because the fire will burn out all the tiny eggs of other parasites clinging to the bricks, so that oysters could spawn on them, free from harmful elements concerned, and that of course facilitate their breeding. Oysters can have a life span of eight to ten years, but those collected after five years of farming are the best for consumption..." (Ye, 1999: 353-354, the authors' translation).

3 It is also worth noting that since oysters are rich in zinc and selenium, they have been seen by many as a natural aphrodisiac.

4 Passenger-cum-freight steamers were already available for waterway transportation in Nanshui in the late Qing Dynasty. The four major routes in service were Nanshui to Jiangmen, Nanshui to Shiqi (today's downtown area of Zhongshan City), Nanshui to Guangzhou, and Nanshui to Macau (Zhang, 2009).

5 Lee Kum Sheung and Leung Tong raised a total of three sons and two daughters, namely Lee Shiu Wing (eldest son), Lee Shiu Dang (second son), Lee Ngan Oi (eldest daughter), Lee Shiu Nan (third son), and Lee Ngan Sau (second daughter). Lee Ngan Sau was their only child that born in Macau, others all hailed from Nanshui.

Young Lee Man Tat and his parents. The sickly "Kin Tai" who once worried his parents so much had grown up as strong as an ox.

GROWING UP

In fact I had a brother and a sister before me, but both of them died very young. My mother was so worried about my viability since her third pregnancy, so much so that I was named "Lee Kin Tai" before I was born. It literally means "watch (him) grow big". However, such a name drew often ridicule from my classmates: "Wow you are really big, and I am small!" That's why when I was in the fifth grade, my mother renamed me Lee Man Tat.

Lee Man Tat

INTRODUCTION

In a longish period between the late Qing Dynasty and Second World War, the three big cities in the Pearl River Delta, namely Guangzhou (abbreviated as Yue), Hong Kong (Xiang Gang) and Macau (Ao Men) had usually been referred to as "Yue-Gang-Ao" by the locals collectively. Apart from closely connected to one another commercially, they shared similarities in various aspects of history, culture, language and custom. Nevertheless, in the eyes of the contemporary Cantonese people, the big three never really enjoyed an equal status. There was the old phrase of "Guangzhou city, Hong Kong district, and Macau street" which reflected the perception of general public towards these three: Guangzhou being the most sizable and developed amongst the three, making it deserved to be regarded as a city; Hong Kong, less impressive in terms of size, yet big enough to be regarded as a prosperous "place" or "district"; while Macau was the smallest and consequently be degraded as just a "street".

However, compared to Nanshui of the Xiangshan County, Macau certainly was an important port city. Macau had been for a long time since 1574 the only port in Southern China opened to foreign traders, and was therefore a bustling hub of business. It suffered a major blow at the turn of the 20th century when the British occupied Hong Kong in 1841 and turned the island gradually into a successful entrepot. Macau declined quickly thereafter, but remained a knot of international networking for the region. For instance, in May 1878, when Sun Yat-sen headed for Honolulu for the first time at the age of 12 from his hometown of Xiangshan, he needed to first travel to Macau, and then from Macau to Hong Kong, and finally embarked liners in Hong Kong to Honolulu.

The uniqueness of Macau is glaring if placed in a broader historical context that spans over 500 years. For a long period of time, Guangzhou was the only trading port designated by the Chinese authority for foreign trade. However, foreign merchants were not allowed to stay or move around freely in Guangzhou. They could only sojourn in Macau along with the Portuguese, as the Portuguese had previously secured a privilege to keep a "permanent settlement" there since the Ming Dynasty. Simply put, some four to five hundred years ago, all foreigners and their vessels travelled by sea to China had to reach Macau first.(1) This tiny peninsula is the place where Lee Man Tat was born and grew up. His childhood and adolescent experiences in Macau have left an indelible imprint on his memory, and he cherishes this part of memory very much particularly in his old age.

Business in Macau, Two Generations

Lee Kum Sheung lost almost everything in the 1902 conflagration in Nanshui Street, Nanshui. When he subsequently migrated to Macau with his family, it was said that the only valuable he brought along was an octagonal wall clock saved from the fire. In Cantonese, the pronunciation of a clock or bell is the same as "the end", hence bringing along a clock or presenting it as a gift are gestures commonly regarded by the Cantonese as inauspicious and offensive. The clock that survived the big fire, however, had become a symbol of vitality and resilience of Lee Kum Kee, and was therefore treasured by Lee Kum Sheung as a mascot.

When Lee Kum Sheung first arrived in Macau in 1902, he got himself into a fairly difficult situation. The place also produced and farmed oysters, so the residents were totally no strangers to oyster sauce, stirring an even fiercer competition in Macau's oyster sauce industry. For instance, "Óleo de Ostra Veng Sang," a contemporary Macau oyster sauce store founded in the very same year of 1902, was once a more reputable brand than Lee Kum Kee (Lao, 2014: 69).(2) As the first step, Lee Kum Sheung rented a room on Rua de Tarrafeiro to settle his family, and used it as a mini workshop to produce oyster sauce. Without a store, he himself then had to carry his products to the local grocery stores and restaurants for sale.

While Lee Kum Sheung produced oyster sauce in Macau, its raw material — oyster extracts — still came from Nanshui. Lee Kum Sheung entrusted friends in Nanshui to collect oyster extracts on his behalf, and saved them in the large traditional pear-shaped jars. He would then take a ferry or boat to Nanshui, self-paid for everything, brought them back to Macau, and boiled them down to oyster sauce. Such an operational model proved successful, and therefore in 1906, Lee Kum Sheung managed to rent a larger premise at No. 9 Rua da Praia do

Manduco. The store-cum-residential unit substantially improved his family's livelihood in the very early stage of their migration.

Compared to Nanshui, Macau was indeed superior in trade and commercial activities, thus Lee Kum Kee soon achieved a higher turnover. Even an off-season sale in Macau was better than a peak-season sale in Nanshui as far as oyster sauce was concerned. Lee Kum Sheung's decision to change track to Macau was vindicated. His booming oyster sauce business subsequently allowed him to buy the premises next door — No. 11 Rua da Praia do Manduco — as an extended workshop and family residence. Some twenty years later, it was exactly where Lee Man Tat was born, but unfortunately, Lee Kum Sheung had passed away at the time.

Given the expanded space of the No. 9 and No. 11 Rua da Praia do Manduco, Lee Kum Sheung began to request his children to assist in his oyster sauce business like most of the contemporary Chinese parents. Members of the second generation had to work in either the workshop or the store after school, and if somebody had ceased schooling, he/she would have to contribute full time seven days a week to the family business. The toil and thrift of the family paid off, for merely over a few years, and Lee Kum Kee has emerged as a notable brand of oyster sauce in Macau.

By the early 1920s, a wealthier Lee Kum Sheung started searching for premises again, this time with better location. When he finally found an available option on the busiest avenue of Avenida de Almeida Ribeiro (also commonly known as "San Ma Lo"), he seized the opportunity and tried to buy it from the owner. But the owner refused to sell, and so it was initially rented by Lee Kum Kee.(3) Today, it is still there and has been regarded as the "prime/old Lee Kum Kee store" in Macau. Located at No. 611 Avenida de Almeida Ribeiro, the store has amazingly been running for business for nearly a century now.(4)

Incidentally, the old store is situated at the end of Avenida de Almeida Ribeiro, which is squarely opposite the once busiest Ponte 16 (Pier No. 16) of the Inner Harbour. Its superb location not only proved the sensible choice of Lee Kum Sheung, but the status of Lee Kum Kee in Macau's oyster sauce industry by the 1920s. Ever since its completion in 1918, Avenida de Almeida Ribeiro has been the main avenue of Macau Peninsula, while the Inner Harbour was a historic waterfront. There was no ferry terminal in the Outer Harbour until the 1980s. Thus in general, fishing vessels, cargo ships, and ferries plying between Hong Kong, Macau and the mainland had to berth at Inner Harbour. Ponte 16 was once the main pier for opium trade and coolie trade. In 1937, after the government re-issued its gambling franchise, the new franchisee opened a casino in the nearby Central Hotel. Moreover, they run Tai Loy Ferry to serve gamblers flocked in from the Mainland and Hong Kong to Ponte 16, thus prompted prosperity of the Inner harbour region.

Lee Kum Kee, in its earliest stage of development, was no more than a humble family workshop. There was initially no clear division of labour among Lee Kum Sheung and his three sons. As his sons grew up and the business continued to prosper, the dividing line naturally took place at the second generation. From the procurement of raw materials to sauce production and sales, the three sons cooperated to handle different tasks, allowing them to get familiar with their respective duties and certain affairs.

Lee Shiu Wing, the eldest, assisted his father in taking care of the store. Lee Shiu Dang, the second, was in charge of sales and marketing. Lee Shiu Nan, the youngest, always worked with Lee Kum Sheung in the workshop, and accompanied him to Nanshui to collect oyster extracts. Lee Shiu Nan soon acquired expertise needed in making quality oyster sauce, and when he was around 13 or 14 years old, he volunteered to deal with the sourcing in Nanshui alone, so as to save his aged father

The Lee Kum Kee store on Avenida de Almeida Ribeiro, Macau, in the 1930s. Although the flow of people was less spectacular then, the location was already a commercial hub.

the hassle of travelling back and forth. However, Nanshui was then infested with bandits and notoriously insecure. Law and order was on the verge of breaking down in the region, but Lee Shiu Nan needed to pay cash (silver coins) for oyster extracts. For safety reasons, Leung Tong sewed an inner pocket for his son to keep silver coins, and had him covered with a black Canton-silk shirt.(5) Moreover, Lee Shiu Nan had to pretend to be a deaf and dumb throughout the trips. He simply kept his mouth shut in order to avoid any attention from the bandits.

Allegedly, boats plying between Macau and Nanshui were full of bandits. Lee Shiu Nan cautiously avoided taking any bath during the trips, lest exposing the inner pocket. He kept sourcing oyster extracts in this way for around three to four years, always under intensive pressure to avoid robbery and also exhausted by heat. The weary Lee Shiu Nan eventually fell sick after a trip to Nanshui. He was infected with typhoid fever and at one point, couldn't even get off his bed. After this distressful incident, Lee Kum Sheung decided to rid Lee Shiu Nan of the burden. At that time, he had already opened a store at No. 151 Avenida de Almeida Ribeiro, and since it was located exactly by the seafront of Inner Harbour, waterway transportation was conveniently available. Lee Kum Sheung subsequently lifted his sourcing price for oyster extracts, so that he could request friends in Nanshui a step further by sending out oyster extracts to Macau.

Compared to its early days in Nanshui, Lee Kum Kee experienced a significant growth during its first two decades in Macau. This less glamourous oyster sauce business had allowed Lee Kum Sheung to feed his family, own a few properties, and get his three sons married. The Lee family soon expanded in numbers. Lee Kum Sheung had three sons and two daughters, and when it comes to the second generation, Lee Shiu Wing had four sons and four daughters, Lee Shiu Dang two sons and seven daughters, and Lee Shiu Nan two sons and six daughters. By the time Lee Kum Sheung passed away, some of his grandchildren were

The old shop at No. 611, Avenida de Almeida Ribeiro in Macau once presented its brand name in trendy fonts, which duly recorded different periods that Lee Kum Kee had passed through.

just about to be born, including Lee Man Tat.

"Grow Big" Wishes of the Parents

Lee Shiu Nan was the fourth child and the third son of Lee Kum Sheung. He married Chan Choi Kam in the mid-1920s. Elegant and virtuous, Chan Choi Kam came from a wealthy family in Macau. She soon gave birth to a son and then a daughter after the marriage, but unfortunately, both of them died at the age of two due to poor hygienic and medical conditions at the time. The couple was devastated and deeply worried, as they might end up with no offspring to continue the family lineage.

Chan Choi Kam was pregnant again in 1928. The news excited Lee Shiu Nan as well as Lee Kum Sheung. However, Lee Kum Sheung fell ill shortly after and his health deteriorated rapidly. He passed away at the end of the year, without the chance of meeting this grandson. His death saddened the whole family, and when they dealt with his obituary, Lee Man Tat was listed as one of his filial grandsons, even though he was still in his mother's womb. However, the name listed was "Kin Tai" instead of today's "Man Tat", which literally means "watch (him) grow big". It was clearly a name with the blessings of seeing him grow up healthy.

The name Kin Tai definitely reflected Lee Shiu Nan's anxiety about the viability of his unborn baby. Given the previous death of his two toddlers, he simply could not dismiss such a lingering fear. On 5 January 1929, Chan Choi Kam gave birth to a baby boy. The couple took extreme care of their baby since then, as baby Kin Tai was in ill health and sickly. They were very much distraught and feared that a similar tragedy might happen again.

When the little Kin Tai cried, coughed or appeared to have any symptoms of illness, Chan Choi Kam and Lee Shiu Nan would simply

The second generation of Lee Kum Kee: Lee Shiu Nan and his wife Chan Choi Kam. Lee Shiu Nan, like his father Lee Kum Sheung, still frowned; while Chan Choi Kam appeared to force a smile. These might be a reflection of Lee Kum Kee's bumpy road ahead.

become jittery and sleepless, not to mention being able to enjoy a meal. Lee Kin Tai nevertheless displayed a great tenacity in life, miraculously growing up. It seemed to be an auspicious new beginning, as the couple's seven other children after Lee Kin Tai have all grown up healthily.(6)

Lee Kin Tai, the sickly little boy, could only be enrolled in Primary One when he was 8 years old (kindergarten education was far from a common practice then). His early schooling in Macau's Chongshi School might not be a happy one, as his name "Kin Tai" was often teased by his classmates. When his parents decided to transfer him from Chongshi School to Pui Ying School(7) at Primary Five, his took such chance to beg to his parents for a new name, henceforth a "Lee Man Tat" in Pui Ying School. Many years later, in an interview with the authors, the smiley octogenarian briefed us about the origin of his name as well as his present name:

> *In fact I had a brother and a sister before me, but both of them died very young. My mother was so worried about my viability since her third pregnancy, so much so that I was named "Lee Kin Tai" before born. It literally means "watch (him) grow big". However, such a name drew often ridicule from my classmates: "Wow you are really big, and I am small!" That's why when I was in the fifth grade, my mother renamed me Lee Man Tat.*

Today, Lee Man Tat's former schoolmates and relatives no longer remember his previous name of Kin Tai, a name had him becoming a laughing stock of naughty kids. It is however a name that he keeps dearly in his heart, overfilled with blessings and love of his parents upon him.

Lee Man Tat grew up in the 1930s, the decade oyster sauce became a famous specialty of Macau. Oyster sauce stores prospered in streets nearby the Inner Harbour. Apart from supplying to high-end restaurants popped up amidst the booming gambling industry, Macau's oyster sauce products headed for export generally. The main demand of oyster sauce came from the worldwide overseas Chinese communities, specifically from Chinese restaurants in North America. Lee Kum Kee, at that time run by the second generation, was one that heavily relied on North America market. Its premium oyster sauce was by no means a cheap commodity, and therefore consumed more by the financially superior overseas Chinese. Macau had more than ten oyster sauce brands at the time, and the "big four" were Veng Sang, Lee Kum Kee, Hap Sengg Long, and Fok Thai Heng. Both the stores of Veng Sang and Fok Thai Heng were situated on Rua do Visconde Paco de Arcos, while Lee Kum Kee and Hap Sengg Long on Avenida de Almeida Ribeiro. The "big four" were in fact very close to one another geographically, yet they competed with one another.

Lee Kum Kee's strongest competitor, Óleo de Ostra Veng Sang, owned a private oyster farm in Macau since the 1920s. It implied that Veng Sang could firmly control and manage the whole process of oyster sauce production, and therefore ensured its quality. In contrast, Lee Kum Kee insisted on sourcing quality oyster extracts from Nanshui, thus a higher cost on raw materials, and consequently a higher pricing. The second generation however abided by the core principles of Lee Kum Sheung in production, which were the pursuance of high quality and integrity. As Lee Kum Kee equated to quality oyster sauce, it allowed the brand to price higher even in a highly competitive market, meanwhile remained a frontrunner of the industry in Macau.

It is nevertheless worth noting that since Lee Kum Kee oyster sauce was rather expensive, it wasn't anywhere close to an essential condiment on dining tables of ordinary families, but a condiment for chefs at decent restaurants. Lee Kum Kee's trading networks are beyond Nanshui and Macau ever since the 1910s. In terms of regional network, before the Second World War, its oyster sauce was popular amongst wealthy families and high-end restaurants in Jiangmen, Shiqi, Guangzhou, Foshan, Dongguan and other places in the Pearl River Delta. And beyond China, its sales reached out to populated overseas Chinese communities in North America (e.g. San Francisco) and Australia (e.g. Melbourne) via trading agents (i.e. Kam Shan Chong) in Hong Kong and Macau.

Lee Kum Sheung was both an aggressive entrepreneur and a prudent investor. He would always retain certain profits from his thriving oyster sauce business for real estate investments. In addition to premises in Macau, he also started to dabble in Guangzhou's property market in his old age. His outlook and vision on property investment, like his ways of managing the family business, had a far-reaching impact on his future generations.

Lee Man Tat grew up in a comfortably-off family entrenched by his late grandfather Lee Kum Sheung. It seemed that the little boy tried to work hard in school initially, but according to his childhood playmate cum classmate Ho Sio Chong, Lee Man Tat was not really interested in learning "boring" knowledge from textbooks, even though he is smart and has a quick and brilliant mind in analysis. Worse was the scenario of having "boring" teachers who relied on textbooks to teach word by word or sentence by sentence. Lee Man Tat, in such occasion, could not even sit still in his classroom. Ho Sio Chong also highlighted his buddy's sensitivity to numbers as well as to commercial opportunities, ever since his very young age. These are traits that Ho Sio Chong believes have contributed greatly to Lee Man Tat's future success in the business world.

When sharing his childhood memory with the authors, Lee Man Tat seldom mentioned stories or events related to his schooling and study. Instead, memories in his mind were always concerning business,

Lee Man Tat in his youth: spirited and confident.

staffing, or even customers. He described Lee Kum Kee as a small-scale, local and typical family business. By the 1930s, only a few people were actual paid employees of Lee Kum Kee, and all of them belonged to the Lee Clan of Chongli Village, Qibao Township, the same clan and hometown of the Lee family. As it was a big family, when the family business — i.e. Lee Kum Kee — received orders, each family member, including children and women, was obliged to help the production. At that time, they dealt with only two items, namely oyster sauce and shrimp paste. They had no idea or intention to grade their oyster sauce, as they sold only the best quality sauce. Regarding the second generation, Lee Man Tat revealed that they didn't spend much time in schools and got involved in the family business as early as possible. It implies that under such family milieu, Lee Man Tat was naturally nurtured a propensity for business and talking business instead of a passion for academic study.

An anecdote that deeply impressed Lee Man Tat during his childhood and has profoundly set his business ethos is as follows: Shrimp paste needs to be dried under the scorching sun in its production. One day, the on-duty Lee Man Tat was distracted away from the drying yard, and when he returned, he saw Lee Shiu Nan throwing out all the fly-covered pastes from winnowing baskets. His negligence had resulted in exposure of some shrimp pastes to flies from their covering veils. It was indeed his fault, but why the waste? He simply couldn't get it. Nobody knew it anyway. Moreover, as he could see, other people kept their shrimp pastes even after they were occasionally licked by the flies. Lee Shiu Nan's reprimand made him understand that "quality is everything", and that it should not be compromised in any circumstance. Lee Man Tat in fact learnt more from the actual practices and behaviours of Lee Shiu Nan than his words and reprimands, for Lee Shiu Nan was simply a reticent man with very few words.

Lee Kum Kee produces shrimp paste along with oyster sauce. The procurement of freshly-caught shrimps on shore to make shrimp paste was the secret in keeping its quality.

A Tumultuous Adolescence

In 1932, the second generation of the Lee family decided to move Lee Kum Kee from its base in Macau to the rising new international entrepot and trading centre: Hong Kong. It was a far-reaching decision based on the momentous change of overseas markets and trading networks of oyster sauce. The new Lee Kum Kee workshop-cum-store was located at No. 262 of Queen's Road Central in Central, Hong Kong. The two uncles of Lee Man Tat moved to live in Hong Kong thereafter, while his father Lee Shiu Nan spent time in both Hong Kong and Macau, occasionally also had to travel to Nanshui and Guangzhou for sourcing affairs and investments. Nevertheless, Lee Man Tat, together with most of his young siblings and cousins, still lived and studied in Macau.

After completing his primary education in Pui Ying School, Lee Man Tat continued his secondary education to Kwong Tai Middle School, which was then affiliated to the private Canton University. He spent three years there to wrap up his junior high education, and then transferred to Chee Yung High School. Amongst his classmates in Kwong Tai were his childhood buddy Ho Sio Chong and his future soulmate, Choi May Ling. Choi May Ling came from an affluent family in Macau, a daughter of the respected entrepreneur Choi Hac Teng. There was nevertheless no romance at all between the two in the middle school. Lee Man Tat, as mentioned, was obsessed with business ideas throughout his adolescence, and always admitted that he lacked the passion for academic study. On the other hand, Choi May Ling, the belle of the school, never even interacted personally with Lee Man Tat in the class, and had a very vague impression of him. It showed that Lee Man Tat apparently did not excel in class.

Apart from distractions of all those business fantasies, another reason that led to Lee Man Tat's lackluster performance in school was his involvement in the family workshop for production. He volunteered to assist his hard-working father in the workshop, and sometimes it took several hours for a round of boiling and stirring the huge wok of oyster extracts, thus missed the class. Compared to other teenagers who missed class due to affections or sorrows similar to "young Werther", Lee Man Tat seemed to deliberately do so for a completely different reason.

School years in youth time are for many their best part of memory, since it is in general a period of colourful and carefree life with big dreams. But to Lee Man Tat, school life passed with little sparkle left in his memory: "In fact, I didn't learn much from the schools. Helping my father at the workshop, though, did allow me to acquire the whole set of skills in producing oyster sauce." Ho Sio Chong, a very close observer of young Lee Man Tat, also commented that "Lee Man Tat was pretty lazy at school and couldn't catch up well with others. But he is such a self-motivated guy who always strives to be the best in business. He is totally obsessed with business."

A broader context of young Lee Man Tat's life in Macau is the Second World War, or more specifically the Sino-Japanese War and the Pacific War, which spanned almost in parallel with his teenage years. When the Japanese finally invaded China in full scale after years of provocations and intrusions in July 1937, they were confident in subjugating China over a mere three months. It was, with the benefit of hindsight, a fatal miscalculation. While the Chinese army and civilians' weaponry were far lagged behind the well-equipped Japanese army, a united spirit and a nationwide mobilisation against the invasion had dragged Japanese military in the vast mainland for years. In order to sustain the war with abundant resources from Southeast Asia and the Pacific region, Japanese military attacked Pearl Harbour of the United States in December 1941, and subsequently invaded and occupied Hong Kong, the Philippines, Malaya and Singapore (Guo, 1979).

The tiny Portuguese colony of Macau had played a special and subtle role in the Second World War. While cities and places surrounding Macau fell into the hands of the Japanese one by one, Macau was spared the trample of Japanese troops due to Portuguese government's wartime declaration of neutrality. Under a facade of peace, Macau had become the only "isolated island of peace" in the Southern China. Yet in reality, Macau was under an ambiguous and indirect control of the Japanese military, and it dared not get rid of the latter's influence. Threats posed by the Japanese military had quickly turned Lee Man Tat into a more mature young man, as he needed to worry about his precarious life as well as his family's safety now. Leung Yin Ming, Principal of the Chongshi School, was assassinated by Japanese agents in Macau in December 1942 for his patriotic anti-Japanese activities,(8) and the primary school happened to be the one Lee Man Tat first enrolled. It is an unforgettable incident in his life. It has evoked a strong emotion deep in his heart, and strengthened his conviction that to avoid bullying, the only salvation for the nation is to strive very hard for self-improvement. This may well be the reason why Lee Man Tat always does not hesitate to serve the nation at his own capacity if given the chances.

Survival in Great Peril

Wartime experience is certainly an important chapter of life for Lee Man Tat. During the Second World War, Macau, despite its uneasy social milieu, saw a massive influx of southward refugees as well as the adventurous traders and capital from the mainland due to its neutrality status. By the end of 1941, when the British in Hong Kong had also surrendered to the Japanese, Macau as the "isolated island of peace" in the region appeared to be even more unique and glaring. Lee Man Tat's uncles and father were forced to return to Macau from Hong Kong, together with a large number of refugees who first fled to Hong Kong. The population of the tiny Macau soon soared to an unsustainable level.

Macau was both miserable and prosperous in the war. On the one hand, influx of refugees had prompted the government to regulate rice market in order to equalise its supply and demand. But the regulation had led to recurrent famines in Macau, and increasingly, people were starved to death on streets. On the other hand, activities of gambling, financing, prostitution and even smuggling reached their heights in the wartime. The rich, the financial speculators, the agents, the traitors and the bandits who fled the battlefield were enjoying a rare luxurious life in Macau (Huang, 1999; Chang, 1999).

Such a conflicting and weird social setting in Macau alerted young Lee Man Tat. When he was out on streets, he realised that unknown dangers might just wait for him somewhere at the corner. Recollecting his days in wartime Macau, Lee Man Tat once mentioned a notorious case of cannibalism that happened in the vicinity of his residence, and that he was one of the witnesses:

> We lived on Rua da Praia do Manduco, and one day, I saw a man carrying a bunch of entrails to the wet market. He got caught and people surprisingly realised that he had slaughtered a fat lady in the ruined old houses on Rua de Inácio Baptista, which was just behind Rua da Praia do Manduco. Of course, they would target the fattest first. You know, it's easy. You need only two people to hold a gunny sack. You target somebody and then you sack him, knock his head to go unconscious. That's it. Lots of people were killed.

Ho Sio Chong also shared a similar memory of cannibalism with us in the interview. Wartime Macau was a miserable presence to him in many aspects, though as a kid from affluent family, he did not suffer a lot personally. He clearly remembers that most people were simply hungry during the wartime:

> Food was still available in Macau, but it was difficult for us to get and

buy. The Japanese had collected food and grains in the mainland for their military forces as much as possible... Many people were starved to death. You see, a young man seized the bread of a woman which she had just bought in the street, and devoured it immediately on the spot. You couldn't stop him by beating him up, since to him, he would rather die in this way than starve to death. At that time, corpses of people who starved to death were commonly seen on streets, but when the authority carried these corpses to the peak hospital (i.e., Kiang Wu Hospital) for dissection, they usually discovered stomachs that filled with horse grains. Macau used to have horse races and a racecourse, these people were extremely hungry, so they ate even horses' food. But horse grains were simply too rough, people couldn't digest them, so they died in this way.

Given the crisis of food shortage, the chaos behind the superficial peace of Macau, and a seemingly endless war, the second generation of the Lee family discussed and decided to temporarily "break up" as a measure of risk dispersion. It means Lee Shiu Wing, Lee Shiu Dang and Lee Shiu Nan would re-settle their own families in three places respectively. They resorted to lot drawing for the destinations of re-settlement. The three brothers' wives came from Qibao, Guangzhou and Macau respectively, and coincidentally, results of the lot drawing matched their wives' hometown. Lee Shiu Wing and his family therefore moved back to the Lee's ancestral hometown in Qibao, Lee Shiu Dang to Guangzhou, while Lee Shiu Nan stayed in Macau. As Macau is the closest to Hong Kong compared to Qibao and Guangzhou, Lee Shiu Nan had to take the risky responsibility of sustaining their store on Queen's Road Central as well. Lee Shiu Wing unfortunately died of illness in Qibao not long after the re-settlement, at the age of only 40-odd years old.

Years of darkness finally ended in August 1945 following the unconditional surrender of the Japanese to the Allied Powers. The

While Hong Kong was under the Japanese occupation, Macau remained "peaceful" due to the neutrality of Portugal. Refugees from South China thus flocked to Macau, which resulted in a sharp rise in its population. The shortage of food left the poor starving to death.

Chinese won their resistance war against Japan. After all the hardships including horror of cannibalism, Lee Man Tat was overjoyed to see peace returning to the region. It was a right moment to him as he had just wrapped up his junior high school education for the senior high. A schooling up to junior high had been regarded as relatively well-educated at the time, and he simply couldn't wait to join the business world. He decided to quit school for something else, though from today's viewpoint, it may still be a time too early to leave school.

Early Career in Guangzhou

Having suffered from the uncertain chaos of wartime Macau, Lee Man Tat yearned for challenges of a normal society to test his ability. He sought his parents' permission to quit school, and received their blessings. He was more than eager to build a career in forming a business.

For Lee Kum Kee, the three families that were forced to separate during the war had then reunited in Macau. In order to carry on Lee Kum Sheung's legacy, they needed to rebuild the brand to a better form. After much deliberation, the Lee family decided to reorganise Lee Kum Kee's business in Hong Kong and Macau, particularly the former. In short, they transformed Hong Kong Lee Kum Kee into a joint-stock company, and tried to expand its market coverage. Amongst the cousins of Lee Man Tat, Lee Man Kwong, Lee Man Dip, Lee Man On and Lee Man Yong had joined Lee Kum Kee in Hong Kong before the war. By the end of the war, they were naturally more experienced and were therefore in charge of Lee Kum Kee.

Lee Shiu Nan was fully aware of his weaker position in the family as the youngest brother in the second generation. He also realised that Lee Man Tat was surrounded by a number of male and elder cousins. It might not be a good idea, therefore, for Lee Man Tat to jump directly from school to Lee Kum Kee. He believed Lee Man Tat should go beyond the family business for a better opportunity. Based on such consideration, Lee Shiu Nan suggested his eldest son to go north for Guangzhou in 1946, together with two younger boys from the Lee clan. The 17-year-old Lee Man Tat was mainly tasked to handle properties of the Lee family in Guangzhou purchased since his grandfather's time. His primary job was to rent out these properties as well as collect rents, and furthermore, to locally produce oyster sauce in a store located on Dade Road and sell them out.

As mentioned, Lee Kum Sheung commenced his investment in Guangzhou in his old age. He prudently bought some small- and medium-size units for the family. When the second generation took over Lee Kum Kee, they followed his conservative style of investment and always reserved certain proportion of the profits for real estate investment, especially properties in Guangzhou.(9) In the early post-war era, the values of these properties were rising, and so the property market of Guangzhou seemed to be more promising than those of Hong Kong and Macau. That could also be a reason why Lee Shiu Nan, after discussing with his wife about their son's future, sent Lee Man Tat to Guangzhou instead of Hong Kong.

Since Lee Man Tat had previously learnt the whole set of techniques in producing oyster sauce, Lee Shiu Nan advised him to set up a workshop within the store building on Dade Road and produced oyster sauce locally. Lee Man Tat excitedly followed his father's advice and worked on the production of oyster sauce in Guangzhou with his two junior partners. He then delivered his handmade oyster sauce by bicycle personally to big restaurants and Wonton-noodle shops in Guangzhou every day. This small business of his own became very promising after only a few years of toil. It gave him a great feeling of satisfaction, and he is in no doubt proud of this achievement even today.

At the young age of 17 to 20 years old, Lee Man Tat learned to become a versatile businessman. He led the three-men team in Guangzhou in the roles of manager, the manufacturer as well as the salesman, which were the same roles as the three brothers of the second generation respectively. He was able to grasp the gist of his business quickly, allocated resources reasonably, and kept his account books clear. Lee Shiu Nan was impressed and that, in turn, encouraged Lee Man Tat to commit more to the job and dream for a dazzling future.

That dazzling future, however, remained a dream to Lee Man Tat. The Chinese Civil War fought between the Kuomintang and the Communist Party of China resumed after the Second World War. State and society shattered, and businesses ruined by the bloody civil war which brought chaos to China again. The Communist squarely defeated Kuomintang (hereafter KMT, also known as Guomindang) in 1949. Its military — Chinese People's Liberation Army, PLA — headed south and finally entered Guangzhou City on 14 October 1949. In the face of such a drastic regime change, Lee Man Tat reluctantly gave up all he had just built in Guangzhou and returned to Macau.

What Lee Man Tat didn't realise was Choi May Ling, the most important person in his life, was in Guangzhou around the same time he left. Once a classmate of Lee Man Tat, this young lady studied at Lingnan University in 1949. However, the two Macau youths did not meet each other in the turbulent Guangzhou. While Lee Man Tat departed for Macau, Choi May Ling was preparing her northward journey to Harbin. Choi May Ling's father stopped her by sending out one of his sons to fetch her back. It was precisely this twist, in the history, that allowed the two to meet in Macau for a romance.

Sales and delivery on bicycles: the perseverant and confident Lee Man
Tat once relied on his own bicycle to sell oyster sauce in Guangzhou.

Conclusion

A shocking blaze in Nanshui destroyed almost all the properties of Lee Kum Sheung, but such disaster couldn't have his entrepreneurship burned. He was insightful and decisive enough to move Lee Kum Kee from Nanshui to Macau, as he apparently noticed that Macau had gradually developed into a regional intermediary for oyster sauce. The following growth and expansion of Lee Kum Kee vindicated his vision. Shortly after Lee Kum Sheung's death, his three sons again moved their base of oyster sauce production and sales from Macau to Hong Kong by securing a store in Queen's Road Central. Moreover, they successively purchased properties in Guangzhou and maintained a store of Lee Kum Kee on Dade Road. In other words, the second generation had fulfilled the wish as well as the grand market positioning of Lee Kum Sheung in establishing a "Lee Kum Kee Oyster Sauce Store, Guangzhou - Hong Kong - Macau".

Lee Man Tat, a member of the third generation, did not enjoy an easy and eventless childhood and adolescence. Yet fortunately, Lee Man Tat seems to have inherited Lee Kum Sheung's blazing passion for a career, as well as the talent of finding ways out and exploring opportunities in disadvantaged circumstances. Ever since his young age, Lee Man Tat has shown a distinctive character of passive in academic learning but highly aggressive in pursuing business. Indeed, he had achieved quite a success in his earliest business career, but as an individual, he couldn't resist impacts that came from a drastic change of the larger politico-economic circumstances. He had to look for greener pastures elsewhere.

1   Qu Dajun, a renowned scholar in the late Ming and early Qing Dynasty, once mentioned in his classical publication of *Guangdong Xinyu* (1678) that "foreigners began to gather and establish their settlement in the two bays. From then on, vessels flocked to Haojing (note: Macau) instead of all the other bays and harbours in the region." In 1689, Qu Dajun again highlighted the significance of Macau in his poem entitled *Macau*: "Amongst the harbours of Guangzhou, Macau is the busiest." These documentary evidences revealed the eminent status of Macau in the contemporary maritime commerce.

2   The founder of Veng Sang, who also came to Macau in 1902, first opened a dried seafood store in the Inner Harbour, and sold his handmade oyster sauce in the store. It was arguably the first oyster sauce store in the Hong Kong and Macau region. Like Lee Kum Kee, Veng Sang was highly praised for its meticulous attitude in hand-making oyster sauce. The founder of the Republic of China, Sun Yat-sen, was allegedly an aficionado of Veng Sang's oyster sauce. When he stayed in Nanjing (Nanking) after the Revolution of 1911, he couldn't help but requested somebody to purchase Veng Sang's oyster sauce for him in Macau. Veng Sang was even invited by the Macau Portuguese Government in 1934 to participate in the Portuguese Colonial Exposition (Exposição Colonial Portuguesa) held in Porto, Portugal. It was a world's fair organised to display achievements of the Portuguese colonies in Africa and Asia, Veng Sang was therefore truly honoured to be selected as a premium brand that represented Macau (Lao Jiayu, 2014: 70).

3   According to available information, this Lee Kum Kee old store was opened for business sometime between 1920 and 1922. It initially applied planks of wood to shut the store. The store functioned as both the wholesale and retail outlet for oyster sauce, and in the early days, its staff would hang a piece of cloth to shelter the room inside, while placing a large glass jar filled with oyster sauce outside. They allowed customers to buy even a small amount of oyster sauce for regular home use, but they should bring their own bowls. As for the store's current façade and structure, it was built during the 1932 renovation.

4   The address of the store was originally No. 151 Avenida de Almeida Ribeiro. After a re-arrangement of the Avenue's number allocation, it has become No. 611 instead of No. 151.

5   Canton Silk is commonly known as Liangchou or Yunsha. It is an artisan fabric made with traditional dying techniques

that have been around for 400 years in Guangdong. It begins with dip-dyeing the fabric in the solution of dye yam. The fabric will then be covered by a locally-acquired river mud full of specific minerals, and be left under the sun for chemical reactions to take place.

6   The elder brother and sister before Lee Man Tat were Lee Ming Yeung and Lee Kam Yong. As both of them died at the age of two, Lee Man Tat has become the eldest child of Lee Shiu Nan. There are six sisters and one brother (Lee Man Lok) after Lee Man Tat.

7   Pui Ying School was located at No. 44 Avenida do Ouvidor Arriaga, but it longer exists today.

8   Leung Yin Ming was then an active community leader. Apart from being Principal of the Chongshi School, he also acted as President of the Macau Overseas Chinese Education Association, and the head of Kuomintang's branch in Macau (Deng et al., 2011: 141).

9   The second generation of the Lee family preferred to purchase properties in Guangzhou instead of Hong Kong and Macau allegedly because they believed that Hong Kong and Macau were after all colonies of foreign powers, thus in the long run, could not be regarded as safe destinations for investment.

3

The young boss in office: Lee Man Tat managed his business affairs in style.

STARTING BUSINESS FROM CRISES

Returning to Macau from Guangzhou, my father was against me about joining Lee Kum Kee. I then proposed to set up my own business. He agreed, and gave me a retail space located on Avenida de Almeida Ribeiro for such purpose. That's why and how I started my own deal. At that time, my main business was trading. I procured some goods from the local markets of Hong Kong and Macau, and had them sold to Portuguese colonies in East and West Africa...

Lee Man Tat

INTRODUCTION

In any society, individuals may differ in their intelligence, industrious spirit and family backgrounds, but when that society is undergoing a momentous and drastic change, every member will bear the brunt of it. The War of Resistance against Japan between 1937 and 1945 was indisputably a war full of agonies and traumas to the Chinese people. When it was finally over in August 1945 with the victory of China, Chinese people naturally expected a restoration of peace and order to rebuild their homes and communities. Lee Man Tat, encountering a similar expectation, decided to kick off his business immediately after the war. He took heed of his father's advice and headed to the biggest city in Southern China — Guangzhou — in 1946. He managed the properties of the Lee family there, and produced oyster sauce locally for sale.

However, much to the disappointment of the people, political situation in China was deteriorating instead of stabilising after the war. The Chinese Civil War erupted again in the mid-1947, and a series of brutal battles had claimed innumerable lives. It ended with a rout of the Kuomintang (KMT) forces to the Chinese Communist Party and shortly afterward a new regime was declared on 1 October 1949. From the Republic of China to the People's Republic of China, state ideology and policies changed drastically from capitalism to socialism. In the face of such a dramatic socio-political transformation, the little experienced young Lee Man Tat was indecisive about his next step. He nevertheless hastily returned to Macau after receiving message from his father Lee Shiu Nan. It was a simple but momentous decision, as his life trajectory then altered profoundly.

Post-war Challenges

A recapitulation of the major developments of Lee Kum Kee in the 1930s and 1940s may help us to figure out its keys to speedy growth and Lee Shiu Nan's expectation of his children on future business development. Shortly after Lee Kum Sheung's death, his three sons already achieved his ambitious market positioning for Lee Kum Kee. Such positioning entailed a significant presence of Lee Kum Kee enterprise in the three vibrant cities of Southern China, namely, Guangzhou, Hong Kong and Macau. A sensible investment approach practised by Lee Kum Sheung had also been followed by the second generation, in which they consciously reserved a part of profits for property investments in these cities. These real estate investments could in turn strengthen Lee Kum Kee's financial situation, thus creating a virtuous cycle for its finance.

That said, Lee Kum Kee's revenue in those days came substantially from the overseas Chinese markets in North America rather than the Southern China markets in Guangzhou, Hong Kong and Macau. Such a revenue structure largely remained unchanged until the 1990s when Lee Kum Kee began to make aggressive inroads into the Mainland China market. Lee Kum Kee's strength in the more profitable overseas markets from the 1930s contributed greatly to its early success, and such success could mostly be attributed to the high quality of its products. Due to the significant freight costs and other expenses for goods exporting, only high-quality commodities could remain competitive in these markets.

Anecdotal evidence suggested that Lee Kum Kee had become a renowned brand amidst the North American Chinese as early as the 1930s, even though on the face of it, Veng Sang was more glamorous locally than Lee Kum Kee and the other "big four" in Macau. Before the Second World War, one could even find Lee Kum Kee's oyster sauce in the French colony of Tahiti, an island in the South Pacific and was a hot sale there. Note that the only product of Lee Kum Kee at the time, its Premium Oyster Sauce, was already standing out from the crowd.(1) The secret of its success, as Lee Man Tat put it, was that "we don't add even a drop of water in it," but simply made sauce out of a large amount of oyster extracts. Such a manufacturing process implied higher costs and consequently a rather expensive retail price. "Only those overseas Chinese could afford our Premium Oyster Sauce," Lee Man Tat added.(2)

As a matter of fact, Lee Kum Kee established and entrenched its reputation in the overseas Chinese communities prior to reinforcing it in the Greater China region. Its oyster sauce was expensive, but exactly how expensive was it? Lee Man Tat gave a comparison of its price with workers' monthly salary in an interview: "A bottle of (Lee Kum Kee's) oyster sauce was sold for HK$ 1.80 in 1945, while ordinary workers earned around HK$ 10 a month. So, that monthly salary was just enough to buy five bottles of oyster sauce. You see, our oyster sauce was really expensive then... But our quality was really good." (Wu, 2013: 81)

Chang Chin Vai, a long-serving Lee Kum Kee staff based in Macau for over six decades, provided another clue for the contemporary price of Lee Kum Kee's oyster sauce in a magazine's feature: "At that time, (monthly) salaries of ordinary workers were only around MOP 8, but a bottle of (Lee Kum Kee's) oyster sauce would cost you MOP 1.20. People in general simply couldn't afford it..." (Yi, 2014: 77) Although Chang Chin Vai did not specify the exact period of time concerned, according to such comparison, an ordinary working family could buy little more than six bottles of oyster sauce by throwing in their whole month's income.

It is worth mentioning that there was no differentiation in the grade and price of Lee Kum Kee's oyster sauce until the early 1970s when Lee Man Tat, at the helm of the company, gradually made reformative changes. Since then, Lee Kum Kee has developed several categories of oyster sauce with different grades and prices. In addition to its classic "Premium Oyster Sauce," Lee Kum Kee now has the economical "Panda Brand" as well as the "Choy Sun" to meet the diverse needs of market. We will return to this issue again in the following chapters.

When Lee Shiu Nan asked Lee Man Tat to head north to Guangzhou in 1946, he gave his eldest son a rather simple mission: manage Lee family's properties in Guangzhou. However, in time, Lee Shiu Nan was impressed by his son's in-born skill in oyster sauce production and acumen in business development. Indeed, the portfolio of Guangzhou properties at the time could be a powerful evidence of Lee Kum Kee's achievement in the North America markets after it had shifted its base from Macau to Hong Kong.

Given the simple task entrusted by his father, the young Lee Man Tat could have simply enjoyed his happy life as a rent-collector in Guangzhou like other youngsters from better-off families. He was, first of all, far away from his parents, and secondly, dealt with the management of properties and therefore in charge of a significant amount of cash. Instead, Lee Man Tat, obsessed with marking his own threshold in business, tried to build his own deal of oyster sauce in Guangzhou through his craftsmanship and the sales network that he had developed.

If it weren't for the civil war and the subsequent regime change, Lee Man Tat might, very likely, settle in Guangzhou, and make it the base of his career. He would probably seize opportunities in other fields to expand his business in the Mainland on top of oyster sauce. The legendary Lee Man Tat might have crafted a very different chapter of his legend.

Guangzhou was liberated by the Communist forces on 14 October

1949. Great challenges posed by such political and economic upheavals were obviously beyond Lee Man Tat's ability to tackle, as he was just a 20-year-old youth. He did, however, attempt to keep the family properties intact, which were accumulated by the company and his father personally over decades. However, to his dismay, the new Guangzhou authorities simply issued decrees one after another to take possession of private properties by making up various reasons. Property owners had little options but to accept the meagre symbolic compensation in return. It was traumatic to Lee Man Tat in losing the grip on these properties under his charge. Even to this day, getting touched on this issue while in the interview with us, Lee Man Tat was overwhelmed by a sense of profound regret and compunction for failing in his task: "I am so sorry about that. All the properties accumulated by Lee Kum Kee and my father in Guangzhou were gone. Our real estate investments were all in Guangzhou." From then on, Lee Man Tat was skeptical of the Communist Chinese regime until the late 1970s, when Deng Xiaoping finally initiated a comprehensive market-oriented economic reform.

Until then, Lee Man Tat had not been closely associated with the family oyster sauce business. Deeply worried about his son's safety in Guangzhou during those uncertain days of 1949, Lee Shiu Nan instructed his immediate return to Macau, even though it implied a hefty financial loss. Lee Man Tat again duly followed his father's instruction and reluctantly left Guangzhou with much chagrin. What was left behind were not just the family's properties, but his local network of oyster sauce business, a familiar business field that he barely started formally.

Fate is woven together like a cloth, hence one's destiny intertwines with many imponderables. And sometimes, we fight to change our destiny. Instead of joining Lee Kum Kee's management team after his reluctant return from Guangzhou, Lee Man Tat opted to start his own business in

Guangzhou was a very impressive business centre with a myriad of Western-style buildings before 1949. Lee Man Tat strived for the sales of Lee Kum Kee's oyster sauce there shortly after he had left school.

Macau, and subsequently made his first bucket of gold in his twenties. He also met his former classmate Choi May Ling again after his return, even though both of them were sojourners in Guangzhou in 1949. They began their love story and tied the knot in December 1954. The marriage has a profound impact on Lee Man Tat as Choi May Ling, his lifelong soulmate, is inseparable from his myriad achievements thereafter.

Starting Business in Macau

Lee Man Tat returned to Macau in the late 1949 with a deep sense of frustration. Lee Kum Kee had lost all its property investments in Guangzhou, but its foundation in Hong Kong and Macau remained sturdy. It might sound logical then for Lee Shiu Nan to allow his eldest son to join Lee Kum Kee, and worked with other members of the third generation in Hong Kong. But the judicious Lee Shiu Nan's previous concerns still remained and, once again, he advised Lee Man Tat to stay out of Lee Kum Kee, at least for the moment.

Lee Shiu Nan simply could not foresee a brighter future for Lee Man Tat in Lee Kum Kee. In fact, as the youngest brother of the second generation, he might have already experienced the problem before. He was relatively serious, reticent, and tended to keep a low profile. He nevertheless outshined his brothers in the sense that he kept every extra dollar for real estate investments. Such a visionary practice had very much strengthened his own branch's standing particularly when, decades later, the three Lee branches entered their negotiations for a separation.

But if a career in Lee Kum Kee was not his course then, where should Lee Man Tat head for? Macau's economy was not in good shape in the early 1950s. Following the outbreak of Korean War (1950-1953), trade embargo were imposed by the United States and its allies against

Communist China. Macau, a tiny economy with meagre manufacturing industry and functioned primarily as a trade nexus between the outer world and Mainland China, was inevitably hit by the regional tension. Lee Shiu Nan gave his eldest son what appeared to be a non-option at the time, which was to set up his own business. Lee Shiu Nan gave him a shop space located on Avenida de Almeida Ribeiro as the base to start his business. The shop was well-situated at No. 129 Avenida de Almeida Ribeiro,(3) which was close to the prime store of Lee Kum Kee at No. 151 as well as to the busy Ponte 16 of the Inner Harbour. Lee Man Tat moved to live upstairs in the two-storey building together with his younger brother, Lee Man Lok.

But the key question was: exactly which business should he engage in? Lee Man Tat, a mere 21-year-old in 1950, pondered over the choice of his own business. He was not interested in the three major manufacturing industries of Macau at the time, namely firecrackers, matches, and incense manufacturing. Unlike some of his peers from affluent Macau families, he would not try his luck on gambling tables either. The general idea was trading, and Lee Man Tat intended to capitalise on Macau's international trading networks. He acutely noticed the glaring gap of economic and social developments between Macau and other Portuguese colonies in Africa,(4) and that trading between these colonies could actually benefit from preferential duty treatments under the Portuguese policy. He wasted no time to seize such opportunity for a trading venture. Recalling how he started his own business, Lee Man Tat briefed us in this way:

> *Returning to Macau from Guangzhou, my father was against me about joining Lee Kum Kee. I then proposed to set up my own business. He agreed, and gave me a retail space located on Avenida de Almeida Ribeiro for such purpose. That's why and how I started my own deal. At that time, my main business was trading. I procured some goods from the local markets of Hong Kong and Macau, and had them sold to*

Portuguese colonies in Africa, during the 1950s, included Cape Verde, Portuguese Guinea,(5) São Tomé and Príncipe, Angola, and Mozambique. These colonies have all become sovereign states later, but even today, they are still relatively poor and underdeveloped, not to mention the time in 1950s. In terms of size and population, the three tiny colonies in West Africa — Cape Verde, Portuguese Guinea, São Tomé and Príncipe — were rather insignificant. But Angola and Mozambique, which are located in the South and the Southeast Africa respectively, were much larger and more densely populated. Thus, when Macau residents talked about "Portuguese colonies in East and West Africa" before the mid-1970s, they were referring mainly to the two big colonies of Angola and Mozambique. Both colonies were rich in natural resources, yet very much lagged behind in manufacturing and other industrial activities. Such weaknesses in their development model, according to certain analyses, were rooted in politico-economic entanglements in Portugal herself.

The main reason that led to such backwardness was a lack of competition in certain economic sectors of the colonies, due to objections from the powerful enterprises in Portugal, or more precisely in the "Metropolitan Portugal". A notable example in this regard came from the sector of cotton textiles: by the 1950s, only one cotton textile mill was allowed in Angola and Mozambique respectively (Clarence-Smith, 1985: 165). Such interventions from the Metropolitan Portugal had depressed economic vitality and development of colonies concerned as a whole. It also resulted in a strong demand for imported light industrial products and daily necessities amidst the Angolans and Mozambicans, even though they were generally impoverished.

Macau's manufacturing industry was relatively developed compared to all other Portuguese colonies in Asia and Africa. The demands for light industrial products, coupled with the preferential duty arrangement for trades amongst the Portuguese colonies, placed Macau in an advantageous position in this trading network. In 1957, the Portuguese government took a step further by abolishing all duties on trade between different colonies (Clarence-Smith, 1985: 165). It was tantamount to the establishment of a free trade zone for Portuguese colonies, which was similar to the Commonwealth's tax exemption arrangement for its members. Accordingly, Macau's export-oriented processing and manufacturing industries thrived in the 1950s and 1960s. A rapid growth of textiles and garments industry followed in the 1970s and 1980s, but it was due to different reasons.

Lee Man Tat's business acumen and talent were fully demonstrated in his quick grasp of the situation, and of the way to cope with it. In 1950, he made use of his old classmates' network to open a store named "Tat Sang Company" at No. 129 Avenida de Almeida Ribeiro. Main items sold in the store were handbags and suitcases, and the supplies came from his old classmate Leung Yee Man, son of the late Principal of Chongshi School Leung Yin Ming.(6) Leung Yee Man was also selling handbags and suitcases but was then based in Hong Kong instead of Macau. By observing traders who busily moved back and forth between Hong Kong and Macau, Lee Man Tat discovered that a few trading houses in these two cities had started to capitalise on preferential duty treatments enjoyed by Macau to make deals with Portuguese colonies in East and West Africa. They would focus on manufacturing a range of light industrial products in Hong Kong and Macau, then had those exported to Africa via Macau.

In the very first phase of his business venture, Lee Man Tat simply copied such model for profits. He purchased commodities directly in Hong Kong, shipped them to Macau, and from Macau exported them mainly to Angola and Mozambique. He sold a great variety of goods out to Africa in his first year of operation, including suitcases, handbags, garments, umbrellas, rubber shoes, plastic buckets, soaps, and so on. He had no preference in selecting items for sale: the only criterion was their popularity amidst the locals. If something proved to be a hot sale in Angola and Mozambique, he would grab it for them. That's why at one time, he even procured cosmetics from trading companies on D'Aguilar Street of Hong Kong for export.

Interestingly, Lee Man Tat once sold a large number of black umbrellas to Mozambique, without knowing why there was such a high demand. Initially, he thought the locals needed them because of the annoying rainy weather or the scorching sun, but he finally realised that in Mozambique, it was customary for the locals to attend funerals with black umbrellas.

The easy way of import-export trading, however, lasted for less than one year. The Economic Bureau of Portuguese Macau soon found out that Lee Man Tat did not actually manufacture goods for export in Macau. He received warning from the Bureau that he could no longer keep his export licence unless he set up a factory for the production. But why didn't he choose to do so in the first place? Lee Man Tat was frank to admit: "I didn't know how to do it." But when he knew that change was inevitable, he promptly adjusted his mindset and *modus operandi*: "So be it. Let's manufacture ourselves."

Lee Man Tat began exploring the manufacturing process and soon mastered the procedures as well as the sourcing of raw materials and manpower. After securing the premises as factory and the required machinery, he set up "Man Sang Leatherware Factory" in 1952. Remarkably, within a few years, Lee Man Tat managed to expand the production lines and set up a total of six factories. These factories, though pretty small in size, handled the manufacturing of light industrial products and daily necessities, primarily leatherwares, umbrellas and garments. He then exported these *bona fide* Macau

Tat Sang Leatherware Company, located at No. 129, Avenida de Almeida Ribeiro, Macau, was the starting point of Lee Man Tat's entrepreneurial venture.

produced commodities to Portuguese colonies in East and West Africa. Lee Man Tat thus fully demonstrated his superb versatility, vigour as well as efficiency in conducting business: nothing was ever too difficult for him!

Admittedly, these factories were small in scale and merely produced low-grade industrial products, which required only simple skills in manufacturing. But they brought to surface the entrepreneurship of young Lee Man Tat: agility of mind, decisive judgement, and proactive approach. Chang Chin Vai, who has followed Lee Man Tat since 1950 when he was running the Tat Sang Company, recalled that all these factories were doing very well at the time, "business was booming, and two of the six factories in Macau were well-reputed."

Lee Man Tat admitted that he didn't make much profit from Tat Sang Company, but earned handsomely from running factories such as Man Sang Leatherware Factory. He wrapped up his manufacturing and trading business in Macau in 1960,(7) not because of lack of profit or priority to the family business since joining Lee Kum Kee. The reason behind was the fraud committed by his right-hand man, Cheung Sing Man.(8) However, to Lee Man Tat, the most important episode during his early years as a businessman was not of commercial consideration, but the fate of encountering with Choi May Ling again in Macau.

Encountering of Old Classmates

Fate is mystical and elusive. Lee Man Tat and Choi May Ling were both born and raised in Macau, and became classmates at Kwong Tai Middle School. After schooling, they took different paths in life, yet ended up in the same town again. It was fate that brought them together and destined them for a romance. In fact, after the end of Second World War, both of them went to Guangzhou for different reasons, but they never met each other there. It was not until 1951 that they met again in the store of Tat Sang Company run by Lee Man Tat.

Choi May Ling came from a distinctive family of Macau. Her maternal great-grandfather, Peng Yulin, was one of the four prominent imperial officials contributed greatly to the restoration of the Qing Dynasty. (9) Her paternal great-grandfather, Choi Wing Kei was also an imperial official with a conferred title of Prefectural Magistrate. Choi Wing Kei was once the head of Ouning County of Jianning Prefecture, Fujian Province, which is part of today's Jian'ou City. He was later transferred to Taiwan Province, by way of Macau. However, it was around this time that the Qing Empire was defeated by the Japanese in the First Sino-Japanese War, and subsequently ceded Taiwan and Penghu Islands to Japan by the terms of the Treaty of Shimonoseki in 1895. Choi Wing Kei therefore decided to settle down in Macau. Choi May Ling's grandfather, Choi Hok Pang (also known as Choi Sai Hong), was one of the four founders of the Macau Tung Sin Tong Charitable Society. Choi Hok Pang was influential in the sense that he partnered with Ko Ho Ning and Lou Kau in the opium trade.(10) Her father, Choi Hac Teng, was one of the earliest graduates (third batch) of St. Stephen's College in Hong Kong. He was in many aspects a chivalrous man with a progressive mind.(11) His progressiveness was reflected in the fact that, unlike most of his wealthy contemporary, he married only once with the woman he loved throughout his life. Lee Man Tat's father, Lee Shiu Nan, shared the rare mindset. Choi May Ling's mother gave birth to a total of 19 children, but only 11 of them grew to maturity. She is her parents' 17th child, the 10th daughter, and the youngest amongst her female siblings.

During the 1950s, Choi Hac Teng was concurrently the Committee/Board Chairman of Kun Iam Tong (Puji Temple), A-Ma Temple, Lin Fong Temple and Kiang Wu Hospital, which reflected his highly esteemed status in the local society. He was also regarded as the Chinese representative by the then Governor of Macau, and would be seated next to the Governor whenever the Governor held a banquet. In comparison, Lee Shiu Nan, a relatively successful Macau entrepreneur, could hardly match his social prestige. Apart from this, the Choi family was financially powerful, with dozens of real estate properties on hand. Members of the Choi family were well educated, and Choi Hac Teng's children were all graduated from or studying in elite Chinese universities such as the University of Nanking. Choi Hac Teng himself also mingled closely with literati in the region. Memory of a visit by Chang Dai-chien, the prodigious painter, to the Choi family, is still fresh in the minds of the cultural elites in Macau these days. Chang Dai-chien was invited by Choi Hac Teng to his Macau mansion in the early 1949, and his stay there for three and a half months as their guest became "talk of the town" in those days.

Choi Hac Teng's mansion was located at No. 14 & No. 16 Rua da Sé, which was close to the Headquarters of Macau Post. Its Chinese-style courtyards were spacious, and the mansion was dubbed by the locals as "Choi's Big House". The mansion was demolished and reconstructed as "Edifício Wing Kei" in the early 1980s. According to Choi May Ling and his younger brother, Choi Cheong Too, the mansion was a two-storey building, which was around 30,000 square feet in area. It was a three-block structure with two internal courtyards. Choi Hac Teng's second son, Choi Cheong Lwan, got acquainted with Chang Dai-chien in Chongqing, and the two became very close friends. Choi Hac Teng thus extended an invitation to Chang Dai-chien when he visited Hong Kong for a joint-exhibition with Yeh Kung-cho on fan painting in the early 1949.

During his stay, Chang Dai-chien and his family lived in the right front room on the ground floor of the mansion. Every morning, he would have a stroll from the Senado Square to the Ruins of St. Paul's, together with his son Paul and the two black gibbons that he bought from Thailand.(12) The story became an anecdote well-known to Macau

Chang Dai-chien (4th right), Lee Man Tat's father-in-law Choi Hac Teng (5th left) and other members of the Choi family.

artists. On the eve of his departure, Chang Dai-chien gave Choi Hac Teng a not-for-sale item from his painting exhibition as a gift, which was entitled *Autumn Hunting of the Tang People*. He also painted directly on four pieces of white satin cheongsam for Choi Hac Teng's children to thank his hospitality (Jiang, 2012).

While Chang Dai-chien enjoyed his stay in Macau, Choi May Ling, the youngest daughter of Choi Hac Teng, was facing the most critical choice of her life. When she graduated from Kwong Tai Middle School at the age of 18 in 1948, she was bewildered by what kind of life she should be leading herself towards. A friend of her asked for her company to attend an entrance examination for nursing course offered by the Hong Kong Sanatorium & Hospital. She thoughtfully took the examination with her friend and passed, while her friend failed. Choi May Ling was then admitted to the course and studied in Hong Kong Sanatorium & Hospital. Half a year later, when Lingnan University — a prestigious church-funded university in Guangzhou — recruited students in Hong Kong, Choi May Ling applied and, passed once again. She therefore decided to quit the nursing course and moved to Guangzhou in 1949 to pursue her university study.

The year of 1949 was indeed a year of traumatic changes. Nevertheless, the campus of Lingnan University as well as the society at large, was relatively peaceful. Choi May Ling therefore did not uproot herself immediately to return to Macau. However, she soon encountered the "Movement to Resist the United States and Aid Korea" following the outbreak of Korean War in 1950, and the whole campus was up in arms. Choi May Ling, today, still has a vivid memory of her student life during the time:

> *The authorities, at that time, stressed that students must go to the countryside to "eat together, live together, and work together" with the local farmers. I was forced to follow others to the countryside. We*

had group discussion every day, and each of us needed to write and submit an autobiography. The group leader would then make a cross reference of these autobiographies, and go after suspicious people... We couldn't really study. I remember they once confiscated somebody's orchard under the banner of communism. The orchard owner was furious and simply couldn't reconcile himself to the cruel fact, so he sent his men to cut down all the fruit trees at night. The orchard owner was then arrested and condemned to be shot. The authorities requested students like us to witness his execution. Each of us was given a colour flag, and marched to the execution ground in columns. It was a horrific scene, I couldn't help trembling. Shortly after this event, the authorities urged us to join the military, I mean all of us. But what could I do in the military? I thought air force pilot might be a good idea. But I was neither tall nor heavy enough to be a pilot, so they intended to send me to Harbin to learn Russian.

When Choi Hac Teng learned about the actual situation faced by his youngest daughter in Lingnan University, the distraught father immediately sent his seventh son Choi Cheung Woon to Guangzhou to fetch his sister back. Choi Cheong Ming, Choi Cheung Woon's younger brother, recollected that it was in fact Choi Cheung Woon that brought Choi May Ling to Guangzhou for Lingnan's admission in the first place, so their father asked the same Choi Cheung Woon to "bring your sister back to Macau now! It was you that took her to Guangzhou." Choi May Ling therefore hurried back to Macau as soon as she could. She deplored the sudden interruption of her university education, and refused to give up her study. Choi May Ling later enrolled in Macau's Chu Hai College and majored in accounting study. At the same time, she also took up part-time teaching in Escola Tak Meng.

It was in 1951 when Choi May Ling was attracted by the display of handbags of Tat Sang Company while she shopped around on Avenida de Almeida Ribeiro. As she entered the store, she spotted a familiar face: the boss there was his former classmate at Kwong Tai Middle School. It was a reunion six years since Lee Man Tat left Kwong Tai, but the scene was far from romantic as some might imagine. In fact, it was a pretty awkward situation when Lee Man Tat, shy and at a loss for what to do, quietly walked out of the store.

Choi May Ling apparently had left a strong impression on Lee Man Tat while at school. Lee Man Tat recalled that during his time in Kwong Tai, even though Choi May Ling was in the same class, "she had many wooers, she didn't even gaze at me once." His old friend and former classmate Ho Sio Chong also mentioned that Choi May Ling was "really gorgeous, every boy liked her at first glance. She was the top belle in school." As for Lee Man Tat, indeed he was shy, but as Ho Sio Chong put it, Lee Man Tat had also been dubbed "one of the ten most handsome guys in Macau," and was never short of admirers and wooers.

Falling in Love

Lee Man Tat was obsessed by Choi May Ling after the encounter at Tat Sang Company, but he lacked the opportunity to meet her again. Macau was then a small place with closely connected communities; the locals mostly knew each other, and gossips about young men and women spread easily. Choi May Ling had many suitors. One night in 1953, Chan Wing Hei, one of her former classmates and the heir to Kwong Hing Tai Firecracker Manufacturing Company, threw a birthday party for Choi May Ling in the mansion of another former classmate. Most of their former classmates were invited, but Ho Sio Chong and Lee Man Tat were not on the list.

Lee Man Tat felt truly frustrated and fidgeted for not being invited. His buddy, Ho Sio Chong, realised that he was infatuated by Choi May Ling. Ho Sio Chong's reaction was straightforward and simple:

he cheekily accompanied Lee Man Tat to the party, uninvited. Hence unexpectedly, Choi May Ling met Lee Man Tat again in the party, this time with a vaguely different feeling. But to Lee Man Tat, the feeling was more than ambiguous. Choi May Ling was "The One", period. Recalling his burning passion towards Choi May Ling at the time, the aged Lee Man Tat remained excited: "After the reunion, there was no room for others. She was the girl of my life, no doubt about that."

Nevertheless, their love affair was not at all smooth sailing. After the birthday party, Lee Man Tat decided to visit Choi May Ling at the Choi's mansion at No. 14 & No. 16 of Rua da Sé every afternoon after he finished work in Tat Sang Company. But Choi May Ling honestly told him that she already had a boyfriend and asked Lee Man Tat to stop coming after her, lest he "be beaten up." Lee Man Tat ignored such tender warning and was strong-headed in visiting her every evening after work. He came and waited inside the mansion even when Choi May Ling was not in. "I was moved by his perseverance," Choi May Ling recalled. Lee Man Tat later moved a step forward by picking up Choi May Ling every weekday at Escola Tak Meng, the school which she worked as a part-time teacher, even though he was teased by students of the school. Meanwhile, Choi May Ling's boyfriend was eager to get engaged with her, but Choi May Ling became increasingly hesitant especially because he was a Mainlander. Still haunted by her highly unpleasant experience of a commune-like lifestyle of "eat together, live together, and work together" during her university days in Guangzhou, Choi May Ling refused to follow him to the Mainland, hence the end of their love story.

On the other hand, Lee Man Tat's sincerity as well as his generous and expansive nature, gradually captured the favour of those "aunties and old ladies" in the Choi's mansion. Choi May Ling's grandfather, Choi Hok Pang, had three wives and many children. When these aunties of Choi May Ling got married, they married to the men who practised polygamy as well. Polygamous families were often plagued by domestic politics, and when these aunties were upset by conflicts in their families, Choi Hac Teng always welcomed his sisters to return home with open arms. The Choi's mansion was therefore not only a spacious house, but a populated home packed with extended family members. In Choi May Ling's recollection, Lee Man Tat was — and is — a guy who really knows how to socialise. He would bring with him a large pack of Swiss candies every time when he visited the Choi's mansion for everyone to enjoy. Lee Man Tat was therefore popular among the old ladies in the mansion, and the ninth grandaunt of Choi May Ling as well as her mother were particularly fond of him.

There was then another suitor of Choi May Ling whose name was Cho Man Tat. He visited the Choi's mansion one day and was allowed in by the doorkeeper. It was dim in the parlour, and before light was turned on, Choi May Ling's mother went out to ask: "Who is it?" "I am Man Tat." "Oh, sit down please." However, when the parlour was lit up, she realised that the guy was not Lee Man Tat. "Well, May Ling is not at home. She is out shopping." But in fact, she knew her daughter was in the room. Choi May Ling naturally would not come out, and her mother's preference was crystal clear to her.

Choi May Ling and Lee Man Tat finally became lovers and dated frequently. While dating, Choi May Ling noted that when they were wandering on streets, Lee Man Tat often deliberately took her to the ordinary restaurants rather than the high-end ones for meals, and covertly observed her reactions. But after an enduring marriage of over sixty years, Lee Man Tat should have felt futile in his pointless attempts to "test" Choi May Ling. As it turns out, Choi May Ling is far more frugal than Lee Man Tat in housekeeping.

The Choi family gave Lee Man Tat a big helping hand in his early career of business. Although the two families were not equal match in terms

Dating Choi May Ling and snuggling up to each other.

of social status and wealth, Choi Hac Teng never stopped his youngest daughter to date with Lee Man Tat. According to Choi Cheong Too, soon after they became lovers and met each other's parents, Choi Hac Teng asked about the nature of Lee Man Tat's business. Lee Man Tat mentioned that he was trading in leatherware, adding that he would soon manufacture goods locally for African markets. Choi Hac Teng was quick to introduce Monteiro, then Deputy Director of the Portuguese Macau's Economic Bureau, to Lee Man Tat. The Bureau's Director was Pedro José Lobo, with Monteiro as his deputy. Monteiro was well-acquainted with Choi Hac Teng because his mistress rented a room in the Choi's mansion. Lee Man Tat admitted in his interviews with us that Monteiro did facilitate his business dealings at the time.

Choi Hac Teng never conducted a background check of the Lee family. Instead, it was Lee Shiu Nan that sought help to do so. Choi May Ling only learned about that after the marriage when Lee Shiu Nan casually mentioned on one occasion that her family owned more than a hundred properties. Nevertheless, Lee Shiu Nan did not care about her family's wealth but her upbringing. Lee Shiu Nan was subsequently satisfied with her rather "simple" family background. Despite the fact that Choi May Ling's grandfather married three wives and her uncles likewise practised polygamy, her father abided by his marriage vow to devote to her wife. Lee Shiu Nan was glad that their viewpoints on marriage were similar.

Action speaks louder than words. To the couple, examples led by their parents have profoundly influenced and set the tone for their marriage. Lee Man Tat and Choi May Ling may have quarrels occasionally, but they were never being unfaithful to each other. Over the years, Lee Man Tat also stresses the importance of "family harmony" and its causal relation with one's personal happiness in his discourse and interviews on formal occasions. He is eager to share with others that family harmony brings happiness in life and sustained development for family businesses. Indeed the rationale sounds simple, but practising this for life may not be so. That is also why so many family businesses experience their twist of fate when failing the practice. On 9 April 2008, when Lee Man Tat was inducted into Babson College's Academy of Distinguished Entrepreneurs in the United States, he again highlighted this point in his speech: "Lee Kum Kee is a family business and, in my view, the key to sustaining a family business, and "Family harmony breeds prosperity" as the Chinese saying goes."

When the Lee Family subsequently crafted its Family Constitution, Lee Man Tat wanted to introduce the "Three Forbidden Rules" into it, namely, NO late marriage, NO extra-marital affair, and NO divorce. However, members of the Family Council could not reach a consensus on the objection to late marriage, hence the "Three Forbidden Rules" were not incorporated into the Family Constitution. Nevertheless, Lee Man Tat strictly disciplines himself with the three NOs, and requests these upon his children. Such a strong emphasis on the values of family and marriage is undoubtedly related to his impeccable marriage life with Choi May Ling, which they have enjoyed for over sixty years and still goes on.

Happy Marriage

Ever since the couple's re-encounter in Tat Sang Company in 1951, they fell in love and by the end of 1953, confirmed each other as their desired lifelong partner. Lee Man Tat didn't mention how he proposed to Choi May Ling and how romantic it was, but he certainly made the solemn commitment of keeping their marriage forever, for better or worse, and for richer or poorer. Time passed and sixty years later, when Lee Man Tat delivered his speech at the 60th wedding anniversary cum 85th birthday banquet of the couple on 23 December 2014, he gazed at Choi May Ling on stage soulfully, and said her partnership over the past sixty years "is my greatest gift from heaven above and leaves my life with no regret."

Lee Man Tat's parents were more than happy to learn about their decision to get married. Choi May Ling did not only come from a prestigious family, but was well-educated and virtuous. Likewise, Choi May Ling's parents were equally excited about their marriage, as Lee Man Tat was such a righteous, capable and self-motivated young man. The two families were at ease to various arrangements of their wedding, with the common intention of making it the most joyous and memorable occasion for the couple.

After a series of pre-marriage customary procedures, the two families decided to set their wedding date in the second month of the lunar calendar in 1954. However, Lee Man Tat's ailing second uncle, Lee Shiu Dang, unexpectedly passed away on the 7th day of the first month. The Lee family mourned over his death and changed the wedding schedule. It was deferred to 23 December 1954.

On the day of wedding, Lee Man Tat and Choi May Ling first registered in Hong Kong under the witness of their friends and relatives. They then returned to Macau for their wedding ceremony and banquet at Hotel Central, the best hotel in Macau at that time. Choi May Ling's younger brother, Choi Cheong Too, was the best man of the day. What's embedded in his memory of the day was a grand and lively wedding, apart from members of the two families, the couple's former classmates in Macau also became focus of the banquet. Inevitably, the newlyweds were teased by friends and relatives for fun and the enjoyment of all.

The grand wedding of Lee Man Tat and Choi May Ling had drawn attention from Hong Kong media, which had reported details of the wedding in their newspapers including photo snaps. Those photos are now naturally faded, but still kept unscathed in Lee Kum Kee's archival repository. But, more importantly, the smiles which came deep from the

The wedding of Lee Man Tat and Choi May Ling in 1954: differed in dress but concerted in soul.

hearts of the newlyweds and their guests were still in living memory.

Today, Lee Man Tat is still proud of this wedding which was organised mainly by himself. He paid meticulous attention to every detail of the event, and was pleased by the approval of his guests. "You know, that was awesome. The whole bunch of (Macau) police and their band had come to entertain." Since both the Choi and the Lee families were well-connected in Macau and Hong Kong, their guests packed the whole Hotel Central, and together, they had made it a rarely seen grand wedding in Macau in the 1950s.

Apart from their marriage registration in Hong Kong and the banquet in Macau, the couple returned to Hong Kong for more banquets in Aberdeen and Yuen Long, to treat their Hong Kong friends, particularly Lee Kum Kee suppliers of oyster extracts who were based in Yuen Long district. Lee Man Tat paid specific attention to them, since after the outbreak of Korean War in 1950, Lee Kum Kee's source of oyster extracts from Nanshui was interrupted due to trade embargo imposed by the United Nations on the Communist Chinese regime. Lee Kum Kee was forced to seek alternative suppliers, and it gradually established a closer tie with oyster farmers in Lau Fau Shan, Yuen Long thereafter. Lee Man Tat was mindful enough of his network building even in his wedding. He seized this opportunity to treat these oyster farmers a feast at Tai Yuen Restaurant in Lau Fau Shan. Such an arrangement again revealed his care and consideration on long-term business relationship.

Lee Man Tat, the rather successful young entrepreneur, finally entered his next stage of life through marrying his soulmate Choi May Ling. Bearing a heavier responsibility as a husband now, he moved forward in the business world even more aggressively, this time with a grandeur perspective.

A joyous group photo of the two families at the wedding, which included family members from various generations.

A Betrayal of Trust

Lee Man Tat focused more on his business since his happy marriage, but soon experienced a major setback due to a betrayal of trust by his right-hand man. He was very much frustrated by the betrayal, but worse was the consequence that he had to finally let go of the business built by his own hands. The man concerned was Cheung Sing Man, who was loosely related to Lee Man Tat by marriage, and had successfully assisted Lee Man Tat in setting up factory after factory over a short period of time. In other words, Cheung Sing Man did contribute significantly to Lee Man Tat's business with Portuguese colonies in Africa. Lee Man Tat however did not seek his help in the first place. It was fate that brought the two together again after the Chinese Civil War.

Cheung Sing Man was older and much more experienced in life than Lee Man Tat. He was once a fairly powerful figure in the government circle. They knew each other during Lee Man Tat's time in Guangzhou, and became close friends. But Lee Man Tat didn't know much about his background and character. When Lee Man Tat left Guangzhou in 1949, he lost contact with Cheung Sing Man.

Lee Man Tat unexpectedly met Cheung Sing Man again in Macau one day in 1951, on the street, where Cheung Sing Man was selling cigarettes at a stall. Lee Man Tat recalled that after the Communist force seized Guangzhou in October 1949, Cheung Sing Man and his wife hurriedly fled to Macau, but he could not find a job there, and his life was miserable. Lee Man Tat knew he was a smart guy with ability, as well as a seasoned former official good at networking. He was therefore an ideal person to complement Lee Man Tat's lack of experience in the field. Apart from that, Cheung Sing Man was also an old friend and a relative. He invited Cheung Sing Man to Tat Sang Company for a detailed discussion, and subsequently offered him a job. It was willingly accepted by the desperate Cheung Sing Man. Lee Man

Tat clearly remembered that he offered him a generous monthly salary of HK$ 50 for the job.(13)

Lee Man Tat was at that time planning to set up his own factory, but as a rookie, he definitely looked for help from Cheung Sing Man. Cheung Sing Man, on the other hand, did not let him down. Man Sang Leatherware Factory was the result of their early cooperation, and Cheung Sing Man won the trust of Lee Man Tat accordingly. Since then, Cheung Sing Man was in charge of the management and operation of the factories, while Lee Man Tat was responsible for raw materials procurement from Hong Kong. As the business flourished and continued to generate substantial profits, Lee Man Tat trusted Cheung Sing Man even more. He introduced all his acquaintances in government agencies to Cheung Sing Man, and let him deal with the networking. He even passed the company's entire finance management to Cheung Sing Man.

Lee Shiu Dang died in early 1954. Since then, Lee Man Tat's cousins asked him back to help out in Lee Kum Kee. Lee Man Tat subsequently settled down in Hong Kong with Choi May Ling after their marriage in December 1954, and formally became a staff of the Hong Kong Lee Kum Kee. Lee Man Tat nevertheless did not wish to give up his Tat Sang Company and the six factories, as these were all his incarnation. Moreover, his job as a Lee Kum Kee's staff in Hong Kong was like "a piece of cake" to him. He therefore came up with a compromised solution: every morning, when back to the office of Lee Kum Kee in Hong Kong, Lee Man Tat would first pick up a long-distance call from Cheung Sing Man, to be briefed about the specific raw materials needed for the day's production. He would then make arrangements accordingly via local calls. Lee Man Tat was simply too efficient in dealing with these mundane tasks, which could be settled with "just a few simple actions by him." After that, freighter would transport the required raw materials to Macau for further distribution to the six factories for production.

Lee Man Tat did not realise that Cheung Sing Man was not honest in financial management. In fact, Cheung Sing Man appropriated a large sum of money from the company's coffer to his own pocket without Lee Man Tat's notice. It was only in 1959 that Chang Chin Vai, at that time stationed in Tat Sang Company, smelt something fishy. He noticed in particular that cheques mailed to the company named Cheung Sing Man as payee.(14) Chang Chin Vai reported these cases to Lee Man Tat, but his boss simply nodded, and said nothing. Chang Chin Vai was baffled by the reaction of Lee Man Tat. Was it because Lee Man Tat still had full confidence in Cheung Sing Man, or simply because he didn't want to upset their relationship as Cheung Sing Man was such a capable manager? Chang Chin Vai did not have the answer.

In 1960, Cheung Sing Man proposed to handover all the business back to Lee Man Tat. However, the company that he proposed to return had almost nothing left but the name. Cheung Sing Man had established his own network via socialising with the local officials, and he no longer needed Lee Man Tat.

Lee Man Tat once considered to take over the mess and continue to run the factories, but he eventually gave up. It was his first own business venture, and he did earn his "first bucket of gold" from there, and the bucket was indeed a large and heavy one. But sadly, the business ended in disarray. Lee Man Tat was indignant for the betrayal and disappointed with his own poor judgement. He thought he had offered a helping hand to Cheung Sing Man in his worst time, and treated Cheung Sing Man well enough, but this was reciprocated by a betrayal of trust instead of loyalty. The lesson had left a strong imprint on Lee Man Tat, so much so that henceforth, he would not easily trust people fully without a very close look.

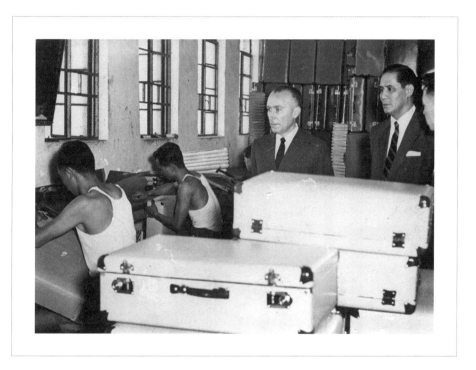

Government officials visited the Man Sang Leatherware Factory to inspect its production.

Since then, Lee Man Tat remains expansive, and may still empower his subordinates when needed. But his most trusted friends, agents and subordinates today are his old friends, old agents and old subordinates that have stood the test of time, since time is the best credit record. When he employs staff, Lee Man Tat would first pay attention to a person's personal attributes and integrity, with ability taken only as secondary consideration.

Decades have passed. At the end of 2012, Lee Man Tat, while interviewed by a magazine and asked about his yardstick and consideration on employing people, he gave a simple but thought provoking answer:

> Q: What is your yardstick in selecting people?
>
> A: One's ability is secondary, the most important quality is a person's "righteousness". If a good person falls short on capability, this can be remedied by engaging someone capable to assist him, so long as he has the magnanimity. But if a person is not righteous at heart, no one is willing to work with him. There are many capable people around, but people with integrity are not easy to come by.
>
> Q: Do you have any religious belief?
>
> A: I don't believe in religion, but I believe in karma. You may not necessarily enjoy a good life by practising good deeds, but if you do evil, you will definitely have a miserable ending. (Wu, 2013: 85)

Lee Man Tat nevertheless did not allow the trauma to stay long in his mind. He soon turned his attention back to Lee Kum Kee and other business opportunities. But the bitter end of Man Sang Leatherware Factory in his career bears a significant implication: it implies the fading out of Macau from his vision. From the 1960s, Macau has become too little to Lee Man Tat. He began to shift his career focus and family life to Hong Kong thoroughly.

Conclusion

In the famous American film of *Forrest Gump* (1994), the lead character Forrest Gump tried to summarise the meaning of life in this fascinating way: "My mom always said life was like a box of chocolates. You never know what you're gonna get." Indeed, one's fate could never be precisely anticipated or resisted, and there are always rise and fall, sweet and bitter moments of life. Lee Man Tat and Choi May Ling both faced a drastic change of their life trajectories in 1949. They returned to Macau, and amazingly met each other again in their hometown, which finally led to a happy marriage.

Cheung Sing Man's life trajectory was likewise drastically changed in 1949. He fled to Macau and received good help from Lee Man Tat, yet he eventually betrayed the trust of this man and caused him damage both financially and emotionally. The betrayal had henceforth affected Lee Man Tat's confidence in people, placing a stronger emphasis on one's integrity. He nevertheless benefitted from the lesson, and by the end of the first decade of his business career, he was better equipped for scaling new heights in his business ventures.

1   Tahiti is the island with most populated Chinese migrants amongst islands in the South Pacific Ocean. These Chinese migrants are mainly Hakka people.

2   The earliest overseas Chinese communities in North America mainly consisted of "Seiyap" (or Sze Yup, meaning "Four Counties") people from the Guangdong Province. Seiyap refers to the four former counties of Xinhui (Sunwui), Taishan (Toisan), Kaiping (Hoiping) and Enping (Yanping) in the Pearl River Delta. There was also an increasing number of Xiangshan (today's Zhongshan) migrants to North America later. For instance, most Xiangshan people headed to Hawaii, while Taishan and Xinhui peoples flocked to San Francisco. In the case of Chicago and New York cities, most of the Chinese there came from Seiyap. This, in fact, was the social foundation of the popularity of Cantonese-style condiments such as oyster sauce amidst the homesick North American Chinese. However, after more than a century, the situation changed significantly: Cantonese are no longer the chunk of overseas Chinese population in North America, instead they are gradually out-numbered by new migrants from the Fujian Province. Moreover, Chinese students and professional migrants originated from provinces of Eastern and Northern China have also increased in numbers over the last few decades.

3   No. 129 was the building number of the shop at the time. The authors do not know its current building number after the authorities re-allocated numbers for buildings on Avenida de Almeida Ribeiro.

4   It is nevertheless noteworthy that since 1 May 1951, following the amendment of its 1933 Constitution and the repeal of its Colonial Act, Portugal is formally no longer a "colonial empire". Portuguese colonies would now be transformed into its "Overseas Provinces". In the case of Macau, despite objections from the Chinese government, Macau was declared an Overseas Province of Portugal in July 1955. Legally speaking, these overseas provinces enjoyed the same status as those in the Metropolitan Portugal.

5   After acquiring its independence from Portugal in 1974, Portuguese Guinea renamed itself as "Guinea-Bissau".

6   Leung Yin Ming was killed by Japanese agents in Macau in the late 1942 due to his deep involvement in the patriotic anti-Japanese activities. It happened during the Second World War when Macau was indirectly controlled by the Japanese.

7   Tat Sang Company, in contrast, continued its

business until the 1970s.

8   This is a pseudonym.

9   The three other prominent imperial officials, in this regard, were Zeng Guofan, Zuo Zongtang and Hu Linyi.

10   Opium was not prohibited in Macau at the time. Before the First Opium War, opium imported to China by the Europeans was mainly via Macau. However, since the cession of Hong Kong Island to the British, Macau's transit status in the opium trade was replaced by Hong Kong (Huang, 1999). The situation changed again when Hong Kong, following the British policy, began to phase out opium trade from 1919; but on the other hand, the Portuguese Macau government didn't do so because it was under fiscal constraint. Opium traders therefore returned to Macau to continue their operations (Booth, 1996).

11   In an article entitled "Choi Hac Teng: The gentleman that I knew" (1990), Wen Dingming noted that Choi Hac Teng hailed from Beiling, Gongbei, which is today a village under Zhuhai City. Choi Hac Teng established Sam Woo Company in Macau before the Anti-Japanese War, which dealt mainly with petroleum business. He also owned the Russian Bakery and Lo Shan Photo Studio. Moreover, he ran the ferry that plied between Macau and Sanbu, Kaiping. During the Anti-Japanese War, Choi Hac Teng, out of his patriotic passion and indignation against the Japanese, accepted request of Beiling villagers and served as their township head. When the Japanese forces captured Gongbei, many Beiling villagers fled to Macau to take refuge. They were all assisted by the generous Choi Hac Teng in Macau. He stopped his ferry service, anchored the ferry on Fai Chi Kei, Macau, and turned it into a temporary shelter for refugee families with children and elders. On the other hand, he allowed refugees who were single to take shelter in the factory building of Sam Woo Company. Choi Hac Teng even distributed food and money to these refugees monthly.

12   It was said that Chang Dai-chien followed a highly routine life during his stay in Macau. He woke up early every morning, and would later walk the two black gibbons with his son. They departed from the Choi's mansion, strolled down Travessa do Roquete to Senado Square, and from St. Dominic's Church turned right to Rua da Palha, and then went straight to the Ruins of St. Paul's. They would then bypass St. Anthony's Church to Luís de Camões Garden for their morning exercise. Such daily schedule was never disrupted.

13   If we refer to the previous note that in 1945,

an ordinary worker earned only around HK$ 10 per month (Wu, 2013: 81), Cheung Sing Man's salary was in fact four times higher than those of ordinary workers.

14   Envelopes of the time were rather thin and semi-transparent, thus allowed Chang Chin Vai to see through the cheques.

Stared in the same future: Lee Man Tat and Charlie Lee.

SOUL SEARCHING

After joining Lee Kum Kee, I surprisingly realised that there was virtually nothing for me to deal with, as we had only one key product, the oyster sauce. Hence throughout this year, I played mahjong for around two hundred days... The International Funeral Parlour, run by Leung Chi Fung, was one of our mahjong game venues back then. One day, when passing through the hall of the funeral parlour, I saw a fine empty coffin. I slept inside to have a try. Lying silently, I couldn't help thinking about my mortal life. What would be my heritage after I left? Or what should I leave behind for the society? And how would people think about me?

Lee Man Tat

INTRODUCTION

International politics witnessed a drastic reshuffle in the early 1950s, particularly with reference to East Asia. It was in the context of such regional transformation that Lee Man Tat, in his twenties, kicked off his own business in Macau. He first opened a leatherware company, then an associated leatherware factory, and from that point expanded to a total of 6 factories in just a few years. He henceforth manufactured goods in Macau for the overseas markets of Portuguese colonies in East and West Africa. Such a rapid expansion of his business proved his efficiency, vigour and most notably, his ambition and the validity of the Chinese common saying, "Never discriminate against a poor young man". Lee Man Tat was particularly envied by the locals in marrying the virtuous and charming Choi May Ling in 1954. The young entrepreneur's grand wedding greatly impressed the locals and made him one of the celebrities in the little "Macau Street".

With Lee Man Tat embarking on a new stage of life in the mid-1950s, both the Lee family and Lee Kum Kee were undergoing changes as well. Lee Shiu Dang, the second uncle of Lee Man Tat who was in charge of Lee Kum Kee's sales and marketing, died in 1954. It was a sorrow moment to the family, and his death affected the management of the brand. Since Lee Man Tat had proven his ability in setting up his own business and performing remarkably well, and had also established himself a family, his cousins invited him to join Lee Kum Kee. They also hoped that he could settle down in Hong Kong, as the new regional developments had posed a greater challenge to Lee Kum Kee. Lee Man Tat accepted their invitation after much deliberation, and most importantly, with the support of his parents. He joined Lee Kum Kee formally in 1955.

Rejoined Lee Kum Kee

As a member of the Lee family, Lee Man Tat certainly had strong feelings towards Lee Kum Kee. It was a family business founded by his grandfather and strengthened by the toil of his father and uncles. In fact, he grew up in an environment that witnessed the daily operation of cooking oyster extracts, packing oyster sauce and selling the products. He was literally growing up together with Lee Kum Kee. During his time in junior high school, Lee Man Tat even acted as an apprentice in the family workshop while assisting his father in oyster sauce production. It is nevertheless worth noting that Lee Kum Kee had at that time moved its base from Macau to Hong Kong. Moreover, Lee Shiu Nan deliberately sent him to Guangzhou after he quitted his senior high school with a clear intention of keeping him away from Lee Kum Kee's business. It resulted in an impression that Lee Man Tat, though a third generation of the family, was an outsider to Lee Kum Kee.

When Lee Shiu Dang passed away in early 1954, reinforcement was needed for the management team of Lee Kum Kee. Lee Man Tat happened to become a good candidate. His talent and entrepreneurship were tested in the advancement of his own venture, and he was just about to get married in the same year. Last but not least, he was the eldest son of Lee Shiu Nan. He was then formally engaged, in the sense that he would then get paid by Lee Kum Kee.

Lee Man Tat's monthly salary was HK$ 250. It seemed to be a substantial amount, but given the socio-economic context of Hong Kong in 1955, he was not paid handsomely. The house prices of Hong Kong, for instance, soared higher than those of Macau since the early 1950s mostly due to influx of refugees. The crazy demand for housing drove up rent prices as well. When Lee Man Tat and Choi May Ling moved to Hong Kong and rented an apartment in Happy Valley, it cost them HK$ 125 per month. The rental therefore already accounted for half of his monthly salary.

Lee Man Tat's first decade in Lee Kum Kee was a time of frustration. He was young, energetic and highly efficient, but then much restricted in realising his full potential within Lee Kum Kee. Such restrictions laid in the fact that Lee Kum Kee remained old-fashioned and conservative in running the business, be it in production, marketing or management. On top of that, members of the family were in general complacent and happy with the status quo of Lee Kum Kee. In contrast, he himself was very aggressive and always striving for something bigger and better. He was confident of his ability and proud of his business acumen. More importantly, Lee Man Tat, in the eyes of his wife, his children and his friends, was a man "crazily" obsessed with business. "Lee Man Tat has no other hobbies apart from doing it."[1] Lee Man Tat could not stand having nothing to do, being obstructed from implementing his idea or unable to manage the business his own way.

In retrospect, Lee Man Tat still grumbled that after joining Lee Kum Kee in Hong Kong, he surprisingly realised that there was virtually nothing to do, as Lee Kum Kee only produced and sold oyster sauce and shrimp paste. Hence throughout his first year in Lee Kum Kee, Lee Man Tat "played mahjong for around two hundred days." He recalled that he could always wrap up tasks concerned quickly, and the "company's business" meant that he needed to go around several Kam Shan Chuang's stores, collecting money and taking orders from them, as well as socialising with the bosses and their staff. Sometimes he also needed to visit the Commerce and Industry Department for certain formalities. In fact, before 1960, Lee Man Tat's Tat Sang Company and the six export-oriented factories in Macau were still in operation. He indirectly managed his personal business in Macau via daily long-distance calls with his right-hand man Cheung Sing Man, and arranged raw materials for the production accordingly in Hong Kong. But to the highly efficient Lee Man Tat, these were just "a piece of cake", and he would wrap up

Lee Kum Kee Oyster Sauce Store at No. 262, Queen's Road, Central, on Hong Kong Island.

things in a short while. In short, he was still pretty bored throughout the day.

One of Lee Man Tat's tasks in Lee Kum Kee was to keep an excellent relationship with Kam Shan Chuang. These were trading agents in Hong Kong that matched tangible needs of overseas Chinese with regional suppliers in Southern China. Lee Kum Kee relied on the overseas Chinese markets in North America heavily; it was therefore crucial to Lee Kum Kee in maintaining good relationships with these trading agents. Since it was Lee Kum Kee — like any other contemporary regional suppliers — that suffered from a disadvantaged position in such unbalanced relationship of power with Kam Shan Chuang, Lee Man Tat soon realised the limitations of Lee Kum Kee's modus operandi. He sharply pointed out that Kam Shan Chuang in fact earned their commissions from two sides of the trading, by raising the original price of a commodity much higher. He was truly unhappy with their overcharge. From then on, Lee Man Tat kept thinking of ways to get rid of Kam Shan Chuang, particularly by establishing Lee Kum Kee's sales and distribution networks independently. However, to his surprise, none of his cousins agreed with him.

Lee Man Tat's cousins were obviously satisfied with the company's status quo of "still making some money," and refused to take risk. But Lee Man Tat thought otherwise. He had confidence in the popularity of Lee Kum Kee's oyster sauce in overseas markets, but if the company's sales network was under the control of a third party, it would be like a chokehold around its neck. Later in the 1960s, he again proposed to his cousins to buy a warehouse as investment. He argued that in order to cope with seasonal changes of demand and improve the company's bargaining capacity, a warehouse was needed to streamline the logistics of production operation. The proposal was, again, turned down. Such conflicts between the conservative and progressive mindsets piled on throughout the years, and finally led to the break-up of the three Lee branches in the early 1970s.

That being said, Lee Man Tat strived to advance the turnover of Lee Kum Kee within the restricted framework, by adopting minor reforms in certain aspects as far as possible. For instance, in the 1960s, eager to make its brand image more prominent, he personally designed a new trademark and packaging for Lee Kum Kee. In 1969, in order to salvage the deteriorating Lee Kum Kee's business in Macau,(2) he also came up with the idea of appointing wholesalers. Apart from retailing, he instructed Chang Chin Vai to promote the large canned oyster sauce directly to Macau's high-end restaurants, such as the Lisboa Chiu Chow Restaurant, as well as prominent bakery shops like the Pastelaria Hin Kee. These restaurants and bakery shops might then split up Lee Kum Kee's oyster sauce into smaller packaging, and sell them out as their own brands. The strategy proved effective in boosting Lee Kum Kee sales in Macau.

Networking via "Yin Nin Group"

The boring and easy tasks in Lee Kum Kee prompted the energetic Lee Man Tat to explore business opportunities elsewhere. He came to Hong Kong in 1955. Given his generous nature and being easy to get along with, he quickly befriended people from various walks of life and established his personal network. The most relevant connection among these was his association with mahjong partners in the business arena, particularly in forming the core circle of "Yin Nin Group", meaning "Prolong Longevity Group".

The concept of "Constant Entrepreneurship" is today one of the core values highlighted by Lee Man Tat. In retrospect, it seemed to be part of his gene, an instinctive impulse without learning, but it may also be forged in his long and exciting business career. The essence of Constant Entrepreneurship is a mindset of never complacent over the status quo, or simply maintaining one's present business, but be constantly innovative to seek breakthrough one after another. And to Lee Man Tat, a breakthrough means an attempt to do things never done before, and never give up.

Lee Man Tat is highly confident in his business acumen, and in practice, he did profit handsomely from his constant investments in different fields. Throughout the 1960s, Lee Man Tat mostly invested through joint-venture in sectors which he was previously unfamiliar with so as to explore new opportunities and hone his business skills. Network capital was the reason why he was able to reach out to these unfamiliar sectors, and amongst his personal networks, Yin Nin Group was a crucial link.

As mentioned, after joining Lee Kum Kee in Hong Kong, Lee Man Tat was tasked with keeping good relations with Kam Shan Chuang. These trading agents, at that time, organised a loosely-affiliated friendship group with their regional suppliers in Hong Kong. The Yin Nin Group was initially nothing more than a socialising network which regularly held feast gatherings for its members. Lee Man Tat once revealed that it was a small group with some 40 members, including himself. The group did not expand its membership throughout the years, thus with passage of time, it gradually lost its aged members, and today only a few were left. Leung Chi Fung, Manager of the International Funeral Parlour, was undoubtedly a key member in Lee Man Tat's Yin Nin Group network. He was a buddy of Lee Man Tat since the 1950s when Lee Man Tat was running his factories in Macau.

Lee Man Tat used to visit Leung Chi Fung's International Funeral Parlour frequently for their mahjong game. The funeral parlour, at that time, was located at No. 41-51 Lockhart Road in Wan Chai. It was re-located to several other places later, and currently settles in Hung Hom, Kowloon with affiliation to the Tung Wah Group of Hospitals.(3) Lee Man Tat said that one day, he went to the International Funeral Parlour again for mahjong game. But when he passed through the hall of the funeral parlour, he saw a fine empty coffin. "I slept inside to have a try. Lying silently, I couldn't help thinking about my mortal life. What would be my heritage after I left? Or what should I leave behind for the society? And how would people think about me?" It is exactly such a thinking and reflection that, years after, as a father, prompted him to take his children with him while attending funerals of friends. The reflective Lee Man Tat is always keen to ask himself what meaningful things could be done in his life, and how to give back to the society? His "taboo-free" strong character and his reluctance to stay put in his comfort zone were also glaring in his act of sleeping in the coffin.

The International Funeral Parlour was not merely a venue for Lee Man Tat's mahjong game. He actually spotted business opportunity in Hong Kong's funeral industry via Leung Chi Fung, and stepped into the field in the 1960s. The modern type of funeral parlours appeared in Hong Kong only after the World War II. They offered professional funeral services in package. For example, when the International Funeral Parlour started its business in Wan Chai in February 1951, its opening advertisement highlighted modern elements of its "contemporary building, standardised facilities, affordable prices, and scientific management". The *Hong Kong Annual 1948* recorded a number of funeral parlours, which were all located in the vicinity of Morrison Hill Road, Wan Chai.(4) Many flower shops and stone tablet shops were then opened on Morrison Hill Road and Wan Chai Road (*Huaqiao Ribao*, 1948). However, by the 1970s, a fierce competition amongst funeral parlours dwindled their number to only three in total, which were subsequently re-located to North Point. This shows that constant entrepreneurship is the key to business success; one either forges ahead or gradually falls behind.

The business opportunity that Lee Man Tat was keen to seize was coffin wood. It is commonly known that in the past, coffin wood in Hong

Kong came from Mainland China, especially Liuzhou. However, since the 1950s, trade embargo imposed on Mainland China had severed the supply chain. Taiwan soon became the alternative source of coffin wood to Hong Kong, particularly the mountainous region of Lishan. Meanwhile, the Kuomintang regime, after its retreat from Mainland China to Taiwan in 1949, was also eager to tap into resources of the overseas Chinese — including Hong Kong Chinese — to boost Taiwan's economic development. In view of this, preferential tax treatments were offered to foreign investors. Taiwan's business environment thus became attractive to Lee Man Tat, and he in fact travelled frequently between Hong Kong and Taiwan during the 1960s to seek opportunities of investment there.

It was also during this period of time that Lee Man Tat began to dabble in the real estate market of Hong Kong. Investment in real estate is kind of a traditional wisdom, and Lee Man Tat, as the son of Lee Shiu Nan and the grandson of Lee Kum Sheung, is always passionate with real estate investment. His real estate investment partners at that time were Leung Chi Fung and Hui Hung. Lee Man Tat took the helm of Lee Kum Kee in 1972. When it came to the 1990s, Lee Kum Kee under his leadership also commenced on real estate investments for the purpose of diversifying the company's assets.

Apart from real estate market, Lee Man Tat also collaborated with others to invest in catering business, and amongst them, "Sha Tin Floating Restaurant" in Sha Tin was the famous one. It was a Chinese restaurant operating on an anchored decorative boat specialising in seafood. The Floating Restaurant had long been a local attraction until 1984, when it was forced to wind up due to reclamation for the New Town of Sha Tin.

Lee Man Tat's investment partner, Hui Hung, or Hui Chi Hung as originally named, was once the right-hand man of Ho Tim,(5) who

Leung Chi Fung, Manager of the International Funeral Parlour, was one of Lee Man Tat's best friends. Lee Man Tat used to play mahjong in his funeral parlour, and once even tried sleeping inside a coffin to reflect upon the meaning of life there. The two posed for a photo during a tour.

was the elder brother of Ho Yin, nicknamed uncrowned "Governor of Macau". Hui Hung at one time also served Liang Cheong(6) in Hong Kong. He was a rare breed of cultured businessman, and particularly excelled in calligraphy. Lee Man Tat was also fond of calligraphy, and therefore felt close to Hui Hung. But a more important bonding in their friendship was the abundant network capital that Hui Hung possessed, which was indeed a tremendous help to Lee Man Tat.

Hui Hung also visited Taiwan frequently with friends of Yin Nin Group such as Lee Man Tat and Leung Chi Fung. His network included many cultural celebrities and political figures in Taiwan, and it was him that connected Lee Man Tat to the famous calligrapher and painter Liu Tai-shi. The friendship between Lee Man Tat and Liu Tai-shi was the most meaningful in the sense that Lee Man Tat later received four scrolls of calligraphy from Liu Tai-shi as a gift, and from all the quotes written on these scrolls, Lee Man Tat was particularly inspired by the four characters of "Si Li Ji Ren" and set the motto as the core value of Lee Kum Kee. Unfortunately, Hui Hung, the man deeply admired by Lee Man Tat, died prematurely in 1974. Lee Man Tat cherished their friendship so much that he never forgot to send Hui Hung's widow greetings and gifts every Chinese New Year and on major festivals.

Lee Man Tat also mentioned in the interview that Hui Hung often took him to massage, and had dinner with him afterwards. Lee Man Tat would habitually request waiters of the restaurant to firstly take out two little dishes of oyster sauce, and taste it personally. He had amazing taste buds, which could immediately discern other brands of oyster sauce from Lee Kum Kee's. According to Hui Hung's younger brother Hui Chi Hing, "Lee Man Tat would be taken aback if oyster sauce used by the restaurant was not from Lee Kum Kee. After returning to the store, he would ask the sales person in charge of that district to follow up the case right away." This anecdote fully indicated that even though Lee Man Tat had got himself busily involved in different businesses in the 1960s, Lee Kum Kee remained his very true love.

From Macau to Hong Kong, Lee Man Tat began his career in Lee Kum Kee by exploring multiple aspects of the business. He then built up his personal network and looked for different business opportunities. The accumulated network capital was vital to him in tackling crises and challenges ahead in the 1960s and 1970s. Before we go through all the exciting stories that happened later, a review of his family life and changes within the Lee family over that period of time may be equally fascinating.

Get Along with Parents

In the eyes of Lee Man Tat, his parents were rather traditional in the sense that Lee Shiu Nan assumed full responsibility as the family's breadwinner, while Chan Choi Kam minded the home. However, his parents were equally unconventional in those days in observing and practising monogamous marriage, i.e. no extramarital affairs. Lee Shiu Nan focused solely on Lee Kum Kee. He was familiar with every step of the oyster sauce production, and paid great attention to the quality of products, regarding it as the life-blood of Lee Kum Kee. Chan Choi Kam took care of their children at home, and continued to care for them after they had all grown up and raised their own kids. Lee Man Tat clearly inherited from his parents his devotion to business and priority for the family. Even though he grew up in the westernised Macau and Hong Kong societies, he had inherited his parents' Chinese way of life as well as Chinese traditional values.

Lee Shiu Nan was the youngest of the three brothers from second generation. Nevertheless, his mastery of the production techniques of oyster sauce made him the mainstay of Lee Kum Kee, and he always attached great importance to quality. After the end of World War II, Lee Kum Kee was fully engaged in post-war recovery work, carrying out company re-organisation to barely modernise its operation. Lee Shiu Nan's role became heavier after the company reform. Lee Kum Kee however remained traditional in its way of production as family workshop, and its financial management was still rudimentary.

For instance, in the case of accounting, Lee Kum Kee, like many other contemporary small and medium-sized companies in Hong Kong, always prepared two different books of account: one for tax filing purposes and the other for detailed records of the company's transactions. When Lee Shiu Dang passed away, Lee Shiu Nan took over the heavy responsibility of book keeping. In fact, Lee Kum Kee's accounts book of the day was never carefully audited by anyone. Lee Kum Kee's revenue as cash would be deposited into a safe, and the key kept by Lee Shiu Nan. The company changed its way of keeping money from the safe to bank accounts only much later.

Although Lee Shiu Nan had stationed in Hong Kong for decades since the 1930s and was always in charge of the workshop of Lee Kum Kee, he did not buy or rent any apartment for accommodation. As a thrifty man, he simply slept on a folding canvas bed in the store of Lee Kum Kee, which was then located at No. 262 Queen's Road Central. When the newly married Lee Man Tat moved to Hong Kong with Choi May Ling in 1955, he used his savings to rent a small apartment on Village Road, Happy Valley, even though rent prices in the area were pretty high. Lee Man Tat's third sister, Lee Kam Yip, lived with the couple in that little apartment for quite a while in order to save rent as she was then studying on Hong Kong Island.

A few years later, when Lee Man Tat wrapped up his trading and manufacturing business in Macau, he used the fortune to purchase an apartment in Sing Woo Road, Happy Valley. He then fetched his mother Chan Choi Kam and his siblings from Macau to Hong Kong to live together in the apartment. The family eventually re-united in Hong

Kong. Lee Man Tat later purchased a more spacious apartment in Blue Pool Road, Happy Valley, and moved the whole family to his new flat, where they could enjoy living in a more comfortable accommodation.

As mentioned in Chapter Two, since 1932, the second generation of Lee Kum Kee had shifted its base from Macau to Hong Kong, while gradually establishing its trading networks amidst the overseas Chinese communities worldwide. It particularly collaborated with those Kam Shan Chuang which specialised in the United States market. A key move then was to open a store at No. 262 Queen's Road Central. It was a rented store, as Lee Kum Kee was still limited in funds at the time. There were other tenants upstairs, and in such a close neighborhood, people interacted frequently and maintained much better relationships than today.

Solicitor Samuel Yu Kwun Chack recalled that when he was just a kid back in the 1960s, his family lived on the second floor of the same building, right above Lee Kum Kee. According to Samuel Yu, Lee Shiu Nan simply slept on a folding canvas bed in the store at night. And every night, around 10 o'clock, he would normally listen to Radio Rediffusion's programme on "ghost stories".[7] As Samuel Yu's father did not subscribe to Radio Rediffusion, he would "eavesdrop" those stories from upstairs. Scary and creepy as they might be, Samuel Yu could not help but keep listening.

Samuel Yu also vividly remembered how Lee Shiu Nan scooped out oyster sauce from the jar repeatedly into his own bottle when he went downstairs to buy a small amount of oyster sauce. He also recalled that on his way home from school, when he smelt the rich flavour of the oyster sauce, he knew he was about to reach home. And the fresh oyster sauce was so seductive that "once I smelt it, my mouth was watering." To Samuel Yu, Lee Shiu Nan was a very serious person, and he would never forget Lee Shiu Nan's chubby body which could often be seen half-naked to the waist in summer months.

By 1968, again according to Samuel Yu, the old building on Queen's Road Central was suddenly listed by the British colonial government as a "dangerous building." It of course greatly affected lives and businesses of all tenants, and so Lee Shiu Nan joined hands with Samuel Yu's father to seek relocation compensation and rehousing from the government, but to no avail. All tenants, including Lee Kum Kee which had already settled there for over 36 years, needed to find their respective new dwellings.(8) It is worth mentioning that Lee Man Tat places a strong emphasis on one's loyalty and trustworthiness when it comes to business relationship. Since he knew Samuel Yu very well, he later engaged him to be consultant solicitor of Lee Kum Kee on intellectual property issues. Samuel Yu had since served in this capacity for a good number of years until his retirement.

After moving out from its old store on Queen's Road Central, Lee Kum Kee had a short stay on Catchick Street, Sai Wan. Its store there resembled the old store in that it was again a store-cum-workshop setting. When Lee Man Tat fully took over Lee Kum Kee's management in 1972, Lee Kum Kee remained as it was in the early post-war period. A year or two later, as a result of the popularity of his new Panda Brand oyster sauce in the United States market, a much larger and more modern plant was urgently needed. Lee Man Tat subsequently acquired a piece of land in Wong Chuk Hang, Hong Kong Island and built a new plant there. It was a new era for Lee Kum Kee in having its own purpose-built factory since then.

Lee Man Tat and his family members shared a great deal of memories of the old store on Queen's Road Central, and Lee Shiu Nan was an inseparable part of such memories. The old store in Hong Kong, Lee Man Tat recalled, was like Lee Kum Sheung's workshop in Macau, which remained a typical family workshop and was stuck to similar

Happy Valley in the 1950s. Lee Man Tat once owned a property on Sing Woo Road, where he settled his mother from Macau to live there together. (Source of photo: Ko Tim Keung)

modus operandi. When the business was good, the whole family would be mobilised; every family member in Hong Kong would be allocated a task, be it the bottling of oyster sauce, the sealing of bottle lids, the pasting of brand stickers, or the cartonning. In those days, staff who did not come from the Lee family were just a few, and they were all originated from the family's ancestral hometown, i.e. Chongli Village of Qibao Town in Xinhui County, Guangdong Province.

While immersed in reminiscence, Lee Man Tat, an octogenarian today, pointed out that within the old building, the store was at front and the workshop at rear. All the employees of Lee Kum Kee were eating, living and entertaining within that little place. They mainly played mahjong as means of entertainment. Cooking of daily meals took place in that limited space as well. In short, "everything had to be conducted within the premises." The place was indeed limited in space, just a little more than 1,000 square feet. As they had a single kitchen and only a big wok, it was impossible to concurrently boil oyster extracts and cook their meal. They had to use the wok in rotation, "when boiling oyster extracts, we had to stop cooking; when cooking, we had to stop boiling oyster extracts." It was certainly not an easy time, but the memory has become an affair to remember for Lee Man Tat decades later.

Oyster sauce is undoubtedly a key element in the childhood memory of Lee Man Tat's children. Lee Man Tat's daughter, Elizabeth Mok, recalled that a glass-frame cabinet was placed at the storefront, and within the cabinet, there were round-shaped large glass urns that kept oyster sauce. People who came to buy oyster sauce in small amount would bring their own containers, and the staff would fill them up with the specific amount of oyster sauce that they bought. Lee Kum Kee indeed also produced canned oyster sauce in one-pound weight, but this was mainly for customers who came from abroad. In other words, it was a specific packaging for export in the early days.

The United States remained Lee Kum Kee's key market over the 1950s and 1960s. When Lee Shiu Nan occasionally learnt about the good news of a big order — which meant an order of 20 wooden boxes of oyster sauce or more — from the United States, he would be very happy and joyously "take out two dollars from his pocket, and ask people to go and buy some extra food items to supplement the meal of the day. He might also drink more during the meal. He has no other hobbies except for running the business and drinking." Choi May Ling recalled.

Lee Shiu Nan was a serious, reticent, and cautious man. He was also thrifty in life. But it was said Lee Shiu Nan could become fiery when scolding people. Chan Choi Kam, on the other hand, was a loving mother at home. The five children of Lee Man Tat all mentioned in their interviews with the authors that they seldom saw Lee Shiu Nan laugh heartily, and they felt emotionally closer to their grandmother. Lee Shiu Nan's old-fashioned frugality could be seen through various ways, for example, even when he was aged over 60 in the 1960s, he insisted to take the tram every day from his residence in Happy Valley to the store of Lee Kum Kee on Queen's Road Central, or to take his breakfast in its vicinity.

Because of his rather heavy-build, Lee Man Tat and Choi May Ling were worried that Lee Shiu Nan might have troubles getting on and off the tram, or even tumble from the tram. They then suggested to him to take taxi instead. But the frugal Lee Shiu Nan refused on the ground that it was too extravagant and wasteful. Elizabeth Mok mentioned that, years later, when he was getting even older, Lee Shiu Nan failed to realise the inflation over the time, thus when they steamed a fresh fish at home and he asked about the price, "we would respond with a much lower fake price to set his heart at ease. Only then, would Grandpa feel that it 'tasted good'; otherwise, he would keep nagging us that we had been cheated by the fishmonger for such an expensive price."

Although Lee Shiu Nan was a man with very few words, Lee Man Tat and his father were much attached to each other. Like many fathers and sons in the world, they seemed to have their own way of communication, which, in this case, was in a tacit and traditional Chinese manner. Moreover, both the father and the son were gastronomes and enjoyed fine dining. Lee Shiu Nan was particularly fond of the dish of "fatty pork meat steamed with salted shrimp paste,"(9) and often enjoyed the dish with rice and liquor. He would even explain to his grandchildren that it was absolutely fine to consume fatty pork meat, as fat would simply turn into water in one's stomach. A vivid impression of Elizabeth Mok of his grandfather is that he loved to drink brandy, and would do so, in small quantity though, while eating his rice. Lee Shiu Nan liked his grandchildren to join him in drinking, even though it meant just a few drops of brandy in a glass of coke.

Charlie Lee, another grandson of Lee Shiu Nan, recalled that when he was still small, he once played around in the old store on Queen's Road Central and felt thirsty. He found "a bottle of water" on the table in the centre of the store, drank the whole bottle right away, and got drunk without realising that this was grandfather's favourite "double distilled baijiu", a Chinese liquor. Charlie Lee, as well as his brothers, all have a good capacity for liquor these days, which sounds like genuinely inherited from their grandfather.

Compared to today's warmer relationship between parents and their children, Lee Man Tat and his parents notably interacted in a subtle way. Although they loved each other dearly, they did not express their strong feelings in words. Lee Man Tat suffered from a period of separation from his family members between Macau and Hong Kong when some had to work or study in different places. Deep in his heart, he fully understood his parents' expectation on him, and they would be pleased only when he really achieved something. He treated and disciplined his children in a similar manner, and hence in the eyes of

The father and son at the dining table: business ethics and values of life are often passed down from generation to generation through occasions of dining and conversation.

his children, he was also, like their grandfather, a serious and strict disciplinarian.

Early Family Days

The most solid foundation of a family business is "the family." Lee Man Tat got married at the end of 1954, and moved to Hong Kong with his wife shortly thereafter. More than a year later, he began to have his children in Hong Kong. His eldest son, Eddy, was born in March 1956. His eldest daughter, Elizabeth, followed in 1957. David, the couple's second son, was born in 1958 and May Mei, the second daughter, in 1960. Unfortunately, May Mei died early. The third and the fourth sons of the couple, Charlie and Sammy, were born in 1961 and 1964 respectively.

Apart from the youngest, Sammy Lee, children of the Lee family were very close in their years of birth, and spent their childhood in the same apartments in Happy Valley. It was therefore a closely connected small family. Decades later, Lee Man Tat's children, as the fourth generation, began to work competently in different positions of the family business. They were in fact the key contributors that propelled the take-off of Lee Kum Kee in the 1980s, led at that time by Lee Man Tat. Choi May Ling once joked in interview that Lee Man Tat kept complaining about the lack of adequate family support in running the family business, "if I could foresee that, I would raise more children for him."

Unlike his father and grandfather, Lee Man Tat and his wife paid special attention to their children's education, Lee Man Tat, however, was too busy to take care of it as he had to concurrently manage the business of Lee Kum Kee and explore business opportunities while minding his own business in Macau. Choi May Ling, therefore, took care of their kids' education single-handedly. Lee Man Tat contributed through leading by example. He consciously guided his children in their vision, interpersonal skills, filial piety and family spirit, as well as nurturing their business skills.

Lee Man Tat's children in the fourth generation shared some highly common observations in regards to their parents' different styles of parenting. Lee Man Tat was a stern father, authoritative and forceful. He was seldom at home, and often returned very late. When he beat and scolded them, he was much ferocious. Lee Man Tat educated them through action rather than words, as he often took them with him while visiting Lee Kum Kee's clients or distributors abroad. A more distinctive feature was that he took funerals more seriously than weddings, and often brought his children with him in attending funerals of his friends. As for their mother Choi May Ling, she was cheerful, optimistic, and always enjoyed good laughs. She kept a simple life for the family, and even when she told the children off, she was relatively "gentle." More importantly, she was impartial towards her children and always treated them equally. On the last score, Choi May Ling stressed that when it came to raising her children, she abided by two rules: "first, no favouritism, absolutely no; second, don't give them too much money to spend!"

The young couple of Lee Man Tat and Choi May Ling, like any others, of course, also had occasional quarrels and discords. Their eldest son, Eddy Lee, remembered that when the family lived on Sing Woo Road, Happy Valley, he was still a little kid then. One day, his parents squabbled over something which he had no idea whatsoever. But his mother was so angry that she picked up a rattan suitcase and headed for her maiden home in Macau. Eddy Lee helped his mother to take the suitcase and followed her to Macau. The mother and son finally returned home in Hong Kong after spending a few days in the Choi's mansion.

Since the family's home was not spacious, the four brothers had to share a room. Lee Man Tat and Choi May Ling creatively designed the room's furniture to suit their needs in that the four beds could be lifted in daytime to make room for their desks. Elizabeth Mok, in contrast, had an exclusive bedroom. She was an obedient girl compared to her four mischievous brothers, and was occasionally bullied by them. For instance, they once scared the living daylight out of her by throwing a string of firecrackers into her room.(10) They would even break up the head and limbs of her doll so much so that she completely lost her mood in playing dolls anymore.

As mentioned, Lee Man Tat was a stern father, and the kids were really scare of his punishment. He wanted his children to go to bed early, and did not allow them to watch too much TV while the kids were crazy about it. Knowing Lee Man Tat would come home late, the kids also stayed up late watching TV paying attention to any sound signifying their father's return. Eddy Lee recalled that he had the best ears, thus when his father pulled out his bunch of keys as he arrived at the door, he would immediately notify others to turn off the TV and go to bed. However, Lee Man Tat was not someone who could easily be deceived. He would find out by touching the top of the TV. He was also against the habit of picky eating amongst his kids, and as his youngest son Sammy Lee recalled, if anyone was spotted picky while having a meal, Lee Man Tat would require him or her to finish the whole dish of food.

The four sons of Lee Man Tat also shared with us their common experience of being beaten up by their father in their childhood. When they did anything wrong, Lee Man Tat would cane them, and they dared not run away. He usually hit them so hard that the cane would sometimes break up. In contrast, Elizabeth Mok was never beaten by her father. She was indeed better behaved, but this might also reflect that Lee Man Tat's mindset of harsher to his sons in penalty. As for Choi May Ling, she might also be upset by her sons occasionally, and pick up the cane to beat them. But as Elizabeth Mok described it, "when

The three generations of the Lee Family: Choi May Ling, Lee Man
Tat, Chan Choi Kam, and the couple's child.

mommy failed to chase up Eddy, she just laughed it off."

Lee Man Tat's children were also impressed by a distinctive practice of their father, which was to bring them together for his business trips. They tripped around the world since young, visiting Lee Kum Kee's clients or distributors in different places. David Lee once laughed about their so-called travelling experience as "accompanying father to visit clients, sitting around and listening to conversations in Taishan (Toisan) dialect." Charlie Lee particularly remembered that when they followed their father to Mexico to visit the family of Cho Loin Chiu, he scratched a luxurious car while riding a bicycle. Cho Loin Chiu was the local distributor of Lee Kum Kee in Mexico. The Cho family has maintained cooperation with Lee Kum Kee since Cho Loin Chiu's father. Cho Loin Chiu's son, Chau Quan Hing, was a young boy back then, but has now inherited his father's business and become Lee Kum Kee's distributor in Mexico. Lee Man Tat purposely guided and nurtured his children through business travels intended to broaden their vision and familiarise them with the sauce and condiment business. Equally important was to connect his children with the next generation of Lee Kum Kee's partners, so that they could establish a personal friendship as early as possible.

A more interesting common recollection of Lee Man Tat's children was about their attending funeral service with their father. They mentioned, in one way or another, how their father deliberately took them to funeral services. Sammy Lee, in the preface of one of his books entitled *The Autopilot Leadership Model* (2016), mentioned that since he was a child, he frequently accompanied Lee Man Tat to attend funerals. "Whenever the master of ceremony read out the deceased's eulogy, it always came across as report card. The achievements of the deceased's life rested in their coffin. Attending funerals has had a great impact on my life. It has always reminded me to think about what I could offer to others, what service and legacy I could leave behind for society." (Lee,

2016: ix)

Apart from the "report card" that reviewed the deceased's life, Lee Man Tat also wanted his children to observe people's emotion and behaviours at the scene of the funeral. How many people were attending the funeral? Who were they? Did they really miss the deceased? And why? Was there any contentious issue or battle on estate sharing amongst the deceased's family members? According to Elizabeth Mok, to this day, Lee Man Tat still attends funerals of his friends with them.

Choi May Ling took care of the children's school education whole-heartedly. Her sons did not excel in their academic study, some even performed poorly. But she did observe that they were quick-witted and had an agile mind. The boys obviously preferred a less constrained learning environment as they were by nature free thinkers. Sammy Lee, for example, was noted by his classmates as "lackadaisical in the classroom, but dynamic in recess" while he was attending St. Joseph's Primary School, but he always had lots of ideas running in his head for earning some little money since childhood. On this score, Lee Man Tat's kids resembled their father very much. Later, they were all arranged to continue their study in Canada and the United States. Their academic performances improved significantly after skipping the spoon-feeding education of Hong Kong, which in fact reveals the importance of educational environment to students.

While living in Happy Valley, Choi May Ling personally drove her children to and from school, including lunch session. The kids would have their lunch at home before returning to school for the afternoon session. She kept this practice to ensure a minimal contact of her children with other school kids lest they would be tarnished. But the practice once led to misunderstanding, as some parents thought that she was an unlicensed shuttle service driver and came to ask for a price for her service. Choi May Ling insisted to do so even when some of her kids had entered secondary school. Eddy Lee, her eldest son, once protested to her: "Mommy, I am already in secondary school! No other mother will fetch her children at secondary school."

Choi May Ling's close friend, Pao Pui Li, mentioned that in order to save some money, Choi May Ling did not order tailor-made school uniform for her children, but re-made them from ready-made clothes. She also kept their school shoes clean and white, as she was habitually a clean and neat person. Moreover, Choi May Ling would frequently make beancurd pudding at home, as Sammy Lee recalled, "she needed semi-porous bags to make beancurd pudding. We helped her to press and squeeze soybeans in the bag and it usually took a toil of 2 hours. Yet when it was finally done, it took us a mere 5 minutes to finish all the beancurd pudding." Choi May Ling's frugality was not all about saving a few dollars to help the family's finance, but a habit, which was also a case of leading by example. The five children of the couple therefore grew up in such a family milieu with both discipline and love. It has become a closely knitted family in which blood is thicker than water and much momentum is generated to propel its vitality.

Shaping Management Ethos and Philosophy

Lee Man Tat frequently visited Taiwan with Hui Hung during the 1950s and 1960s. It was through the personal network of Hui Hung that he luckily met Liu Tai-shi, a prestigious master of Sinology, in Taipei. They found each other congenial, and Liu Tai-shi subsequently wrote down four scrolls of calligraphy for him. Lee Man Tat is very much fond of the gift from Liu Tai-shi, and profoundly enlightened by the words and wisdom therein. Lee Man Tat's motto of "Si Li Ji Ren" came from the scrolls. It means so dearly to him that up to now, the four scrolls of Liu Tai-shi's calligraphy are still hung up in the office of Lee Man Tat at Lee Kum Kee's headquarters in Tai Po, Hong Kong. And in the office, the

The five children grew up while Lee Man Tat's business was on the rise. He thus had more opportunities to take them on trips to visit friends and clients.

scrolls are placed in a position that faces his desk.

Liu Tai-shi (1898-1989) was a native of Xinfeng County, Jiangxi Province. He was a renowned sinologist, calligrapher and painter. Liu Tai-shi came from a scholarly family, and since his childhood, he was purportedly diligent in study and read extensively, hence excelled in the field of sinology from a young age. Liu Tai-shi went to Beijing in 1919, and after an exclusive examination conducted by the then Principal of Peking University, Tsai Yuan-pei, he was admitted into the School of Arts of Peking University.(11) Upon his graduation from Peking University, Liu Tai-shi returned to Ganzhou and served as the head of several counties in the region. After the outbreak of the Anti-Japanese War in 1937, Liu Tai-shi threw himself into the resisting movement and was appointed Secretary of the Ministry of Defence. In 1950, when Liu Tai-shi visited Hong Kong from Taiwan, he interacted with Chang Dai-chien for that short period of time based on mutual appreciation. In 1952, he was invited to the newly found Nanyang University in Singapore and served as its Professor of Sinology for more than six years, where he taught the *Classic of Poetry, Records of the Grand Historia*n and other Chinese classics there. Liu Tai-shi later returned to Taiwan and continued to teach at prestigious universities in Taipei (Fu, 2012). When Lee Man Tat met Liu Tai-shi, it was in his later stage of teaching at universities in Taipei.

What is the wisdom embedded in words of the four scrolls of calligraphy? Contents of the calligraphy are as follows:

The happiest moment in our lives is to keep reading and studying; the most crucial deed in our lives is to educate our children well. A focused mind keeps us calm; a restrained desire makes us healthy.

Never be obsessed with pleasures, as they may lead us to fret. Never reject hard labour, as they will finally bring us peace and health.

Refrain from talking too much to our friends, as words may easily incur grudges. Books could advance our wisdom, and we should therefore urge our offspring to read more.

Cherish our food and clothing, not for the sake of wealth, but for treasuring the blessing itself. It is totally fine for us to pursue fame and fortune, but it should be based solely on our endeavours, rather than on kindness of others.

Fortune usually appreciates a hardworking person. Meanwhile, we should always be cautious over the bad intentions of others.

Having only one confidant in our whole life may well be enough, as confidants are someone to wait for, not to ask for. Self-study should be a life-long process without expiry.

Contentment is a joy in life, non-intervention is the rule of nature.

Stay calm while facing critical incidents in life. Always bear a conviction that sages may still be around even today.

Befriend sages and enjoy their beneficial companion. Compiling articles written by ancient literati is a spectacular deed.

For a person of great ability, it takes time to finally reveal his/her true uniqueness.

An aggressive learner is someone who always feels hunger for knowledge, thus never stops exploring.

Self-cultivation is an end in itself, not for the sake of reputation building; and we should always consider others' interests in work.

One should willingly accept criticism even though he/she has enjoyed a good fame. One should keep learning and studying even after achieving enormous success.

A person with great success should keep himself ambitious, while willing to help or share with others.

Stand firm on our heels and straighten up our back. We could then broaden our perspectives and keep a peaceful mind.

Lee Man Tat was particularly moved by the notion of "considering others' interests (in four Chinese characters of "Si Li Ji Ren") and felt truly enlightened. After a transcendental rumination, he realised that the world is prone to contradictions and conflicts because we usually care solely for our own interests, without considering others'. These include interests of our suppliers, employees, consumers, society, and to a larger extent the natural environment where we live. Such actions of selfishness which focus simply on one's own interests would never help to maintain a lasting and harmonious relationship. However, if we could view the issue from a higher ground, and take into account the well-being and concerns of other parties, they may reciprocally join hands with us in developing business, friendship, and most importantly, trust.

Lee Man Tat has continued to share with his children, relatives, employees and even clients regarding his understanding of the motto since the late 1960s. And, after decades of practice of the value in life, he gradually entrenched "Si Li Ji Ren" as the core value of Lee Kum Kee, with the wish of sustaining the century-old family enterprise even longer. The five children of Lee Man Tat often observed that their father would not only harp on the importance of "Si Li Ji Ren" to them, but would pen it down as well as elaborate it as gist to his clients while negotiating deals with them. Chow Sing Lam, Managing Director of

A huge scroll of calligraphy that highlights the very core value of Lee Man Tat, "Si Li Ji Ren" (considering others' interests), is displayed on a corridor of the Tai Po Headquarters, Hong Kong.

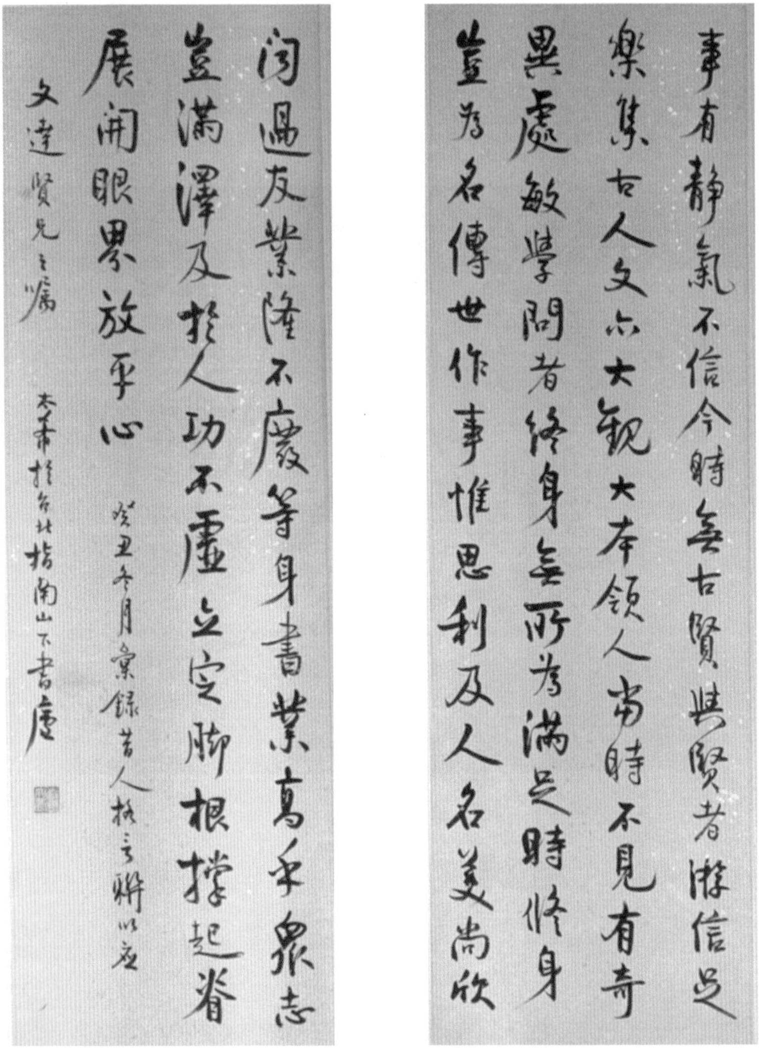

事有靜氣不信今時無古賢其賢者猶信乎
樂集古人文亦大觀大本領人當時不見有奇
異處敏學問者終身無所為滿足時修身
豈為名傳世作事惟思利及人名美尚欣

閱過友誼隆不廢等身書業高千眾志
豈滿澤及於人功不虛立定腳根撐起脊
展開眼界放平心

癸丑季月彙錄古人格言辭以应

文達賢先之囑

志希於廣州指南山下書處

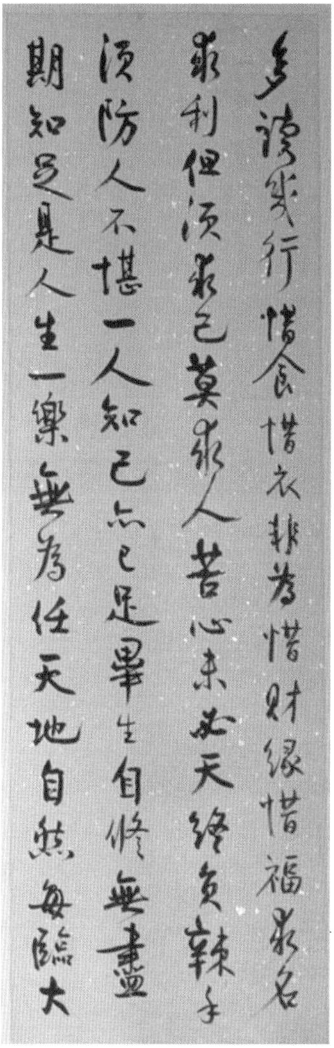

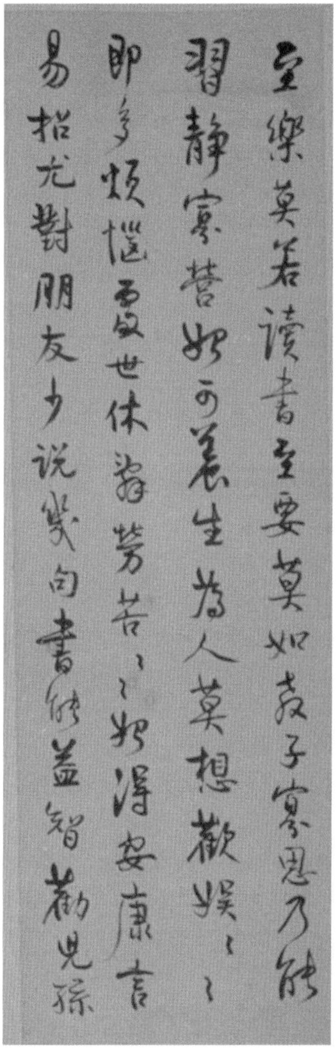

至樂莫若讀書至要莫如教子寧思不缺
習靜寡言如養生為人莫想歡娛
即多煩惱欲世休瑩勞苦得安康言
易招尤對朋友少說笑句書修益智勸兒孫

多讀笑行惜食惜衣非為惜財緣惜福求名
求利但須循禮莫欺人苦心未必天終負辣手
須防人不慊一人知己足畢生自修無盡
期知足是人生一樂無為任天地自然每臨大

The four scrolls of calligraphy were written by Liu Tai-shi, a renowned master of Sinology, as a gift to Lee Man Tat, who was profoundly enlightened by the four characters of "Si Li Ji Ren" and cherished the scrolls very much. They are hung up in his office at the Tai Po Headquarters, Hong Kong.

the Nansen Trading Company, has become Lee Kum Kee's distributor for several decades. He flew with Lee Man Tat to South Korea for the inspection of the local oyster extracts in 1973. On board the flight, he received a paper note of the four Chinese characters of "Si Li Ji Ren" penned by Lee Man Tat. Lee Man Tat said nothing after penning the note, as though the words spoke for themselves.

It is well noted that the couplet of "*Self-cultivation is an end in itself, not for the sake of reputation building; and we should always consider others' interests in work*" first appeared in the Tang Dynasty. It had become a popular couplet in the Chinese society long before the Qing Dynasty. According to Liang Zhangju(12) in his *Yinglian Conghua* (*Notes on Couplet*), the 14 Chinese characters of the couplet in fact came from Yan Zhenqing's model-calligraphy of *Zhengzuowei Tei*. These Chinese characters were separately picked up from *Zhengzuowei Tei* and re-organised by someone as the couplet (Liang, 1987: 2 & 112). It is indeed a concise motto with clear and simple message, and that may well be the reason why Lee Man Tat often thinks that elaboration may not be needed. At the end of 2014, Lee Man Tat, on one public occasion, also expressed that "Si Li Ji Ren are four meaningful Chinese characters. If we follow the meaning of these four characters in life, we could achieve harmony. It is indeed superb that the value could be applied in any field, be it the field of business, family, company or society."(13)

Indeed, "reputation building" is a vanity. The real value of our life is reflected primarily through what we have actually done. But we have to recognise that people are fundamentally motivated by the discernible "interests" that they could ensure from doing things. Sammy Lee and his siblings, i.e. the fourth generation of the Lee family, later attempted to elaborate on the broader concept of "Si Li Ji Ren" more specifically and practically. They therefore initiated the three elements of "Helicopter View", which implies a holistic perspective; "Think from Others' Perspectives"; and "Care about Others' Feelings" as specific approaches to practise the concept.

Sammy Lee generalises the meaning "Si Li Ji Ren" in a simpler way in another publication entitled *The Power of "Si Li Ji Ren": Nine Rules for Life Achievement*. To Sammy Lee, it could simply mean "consider how things could benefit all of us in the first place" (Lee, 2012: 5). In other words, the simple essence of "Si Li Ji Ren" is mutually beneficial and reciprocal. It is nevertheless easier said than done. A lifelong practice of sharing interest with others wholeheartedly is, to say the least, a very difficult task, not to mention a common practice of such philosophy with one's offspring, employees and clients.

Crisis of Oyster Extracts

Lee Man Tat's first decade in Lee Kum Kee was a frustrating experience as most of his reform proposals were simply turned down by his cousins. They preferred to stay in the comfort zone existed and sustained through hard works of their grandfather and fathers, without a sense of crisis. But the very conservative style of management could not ensure a safe passage through the turbulent years of 1950s and 1960s. The sudden outbreak of a crisis related to oyster extracts in 1965 was therefore a destructive hit on the unprepared Lee Kum Kee and almost destroyed the family business. Lee Man Tat responded to the crisis composedly, and impressively tapped into his personal network to deal with troubles. His successful settlement of the crisis once again proved his ability in turning crises into opportunities.

What exactly happened in the crisis that had erupted before Lee Man Tat seized the total control of Lee Kum Kee in 1972? And how severe the impact was on Lee Kum Kee as well as the family? To tell the story, we may need to return to the grand setting of international politics in the early 1950s.

The eruption of the Korean War in 1950 led to trade embargo imposed by the United Nations on the newly-born Chinese Communist regime in Mainland China. The US-led embargo against China prohibited any Hong Kong product which adopted raw materials from Mainland China to be exported to the United States. Under such sanctions, any Hong Kong product exported to the United States must attach a permit issued by the Hong Kong Commerce and Industry Department to prove its "purity" from any Chinese raw materials. It was definitely a serious blow to Hong Kong's trade and industry, which was then export-oriented. As a colonial territory with almost no natural resources, Hong Kong's export commodities at the time could hardly be free of raw materials from the Mainland, and the sauce and condiment industry was no exception. It was in fact an extremely unfair sanction against the small businessmen and factory owners in Hong Kong.

Lee Kum Kee needed a huge amount of oyster extracts in its oyster sauce production, particularly extracts in excellent quality. It needed salt as well to make oyster sauce, but salt was much more easily available than good-quality oyster extracts. At that time, Lee Kum Kee had started to source local oyster extracts from Lau Fau Shan, Yuen Long District, but the local supply of oyster extracts was far from sufficient due to rising demand from its overseas markets after the World War II. A multiple supply of oyster extracts was particularly crucial to Lee Kum Kee in keeping its reputation as the quality producer of oyster sauce.

Like many other contemporary manufacturers in Hong Kong, Lee Kum Kee in 1965 did not actually sever all its connections with the Mainland. Most of its good-quality oyster extracts still came from certain sources in the Mainland. Lee Man Tat's second cousin, Lee Man Tip, was previously in charge of Lee Kum Kee's sales and external affairs. He was a capable and skillful businessman who came from Lee Shiu Wing's family. His approach to tackle the sanctions was no different from the vast majority of other Chinese small businessmen and manufacturers, which was to keep two accounts books in terms of procurement of raw materials. On the other hand, the British colonial government might not strictly supervise the situation as well. Lee Kum Kee's official accounts book for tax filing therefore noted all its sources of raw materials as "Hong Kong," and was able to keep itself out of any trouble by following such practices for years. Lee Man Tip however suffered from a stroke at Tai Tung Restaurant, Central and died at the scene in 1964. When Lee Kum Kee suddenly hit by the crisis of oyster extracts in 1965, Lee Man Tip was no longer there to take command.

It was a day in 1965 when inspectors from the Commerce and Industry Department, as usual, came to Lee Kum Kee for a routine audit of its accounts, checking sources of its raw materials. But a negligence committed by a Lee Kum Kee staff consequently triggered a crisis that almost engulfed Lee Kum Kee. Recalling the event, Lee Man Tat observed that:

> *The company's accountant inadvertently placed the receipts and accounts book on desk, which were related to oyster sauce procurement from Mainland China. The inspectors soon discovered these and yelled: "What? Procurement from China again?" They said "no way," and ordered to lock the door. They then called over some of their more senior staff to come and check, searched through our accounts books, and took away a full 20 boxes of our accounts books and other related materials. Thereafter, the Commerce and Industry Department suspended our licence, and we could no longer export to the United States.*

It was apparent that within the Lee family, the tactful Lee Man Tat was the best man to deal with such crisis. He was given the mandate, and he therefore began to deliberate on the appropriate strategy and action to respond. His personal network, at that time, lacked persons that were really influential to the British colonial administration. He first sought help from the Chairman of Kam Shan Chuang Association, Lee Yu Chee. Lee Yu Chee was a man very good in networking and knew many people, but he failed to find a way out for Lee Man Tat. Lee Yu Chee nevertheless suggested Lee Man Tat to approach Leung Chi Fung, who was Manager of the International Funeral Parlour and a member of Yin Nin Group. Leung Chi Fung subsequently reminded Lee Man Tat that their common friend, Hui Hung, knew Richard Charles Lee. Richard Charles Lee's cousin, Lee Quo-wei,(14) was then a committee member in the Commerce and Industry Department. Networks of these influential figures might well be a way for solution, as they could mediate for Lee Kum Kee within the administration.

Lee Man Tat approached Hui Hung and finally confirmed an appointment with Lee Quo-wei. He visited Lee Quo-wei's office with Leung Chi Fung, and as he recalled, the two male secretaries of Lee Quo-wei were also at the meeting. Lee Man Tat was well prepared with a "reasonable explanation" to brief Lee Quo-wei that "Lee Kum Kee did buy some oyster extracts from Mainland China, but these were only for the production of oyster sauce exporting to Canada, not to the United States." Trade embargo of the day was in fact limited to the United States market, while the Canadian market was not restricted. Bearing a tacit understanding in mind, Lee Quo-wei simply responded with a few words: "OK, now we have learnt your case. Kindly go back and wait for the result." Indeed, more than a week later, Lee Man Tat received a letter from the Commerce and Industry Department, asking him to visit the department again for questioning.

The licensee of Lee Kum Kee, at that time, was Lee Man Kwong. He was the first cousin of Lee Man Tat and came from Lee Shiu Wing's branch. Lee Man Kwong was scared by the crisis, with a deep fear of being arrested and prosecuted, thus he dared not visit the government officials. Lee Man Tip was a quick-witted and capable member of the family, but he unfortunately passed away before the crisis. As a result,

Lee Man Tat became the only hope for the family to cope with the crisis, even though he was still relatively young.

Lee Man Tat recalled that since then, Lee Kum Kee's licensee had been changed from Lee Man Kwong to Lee Man Tat. Lee Man Tat was then legally bound to interact with the government, and he had to bear the legal consequences from then on. As Lee Man Tat did not want to fight an unprepared battle, he inquired information from the inner circle in advance, and get everything well settled before he finally stepped into the Commerce and Industry Department. As expected, Lawrence Mills, the then Deputy Director of Commerce and Industry, asked him to sign a document and immediately to declare "no more purchase of Chinese raw materials henceforth." But apart from that, the department returned its export quota to Lee Kum Kee, and imposed no fines. In fact, it even raised the export quota for Lee Kum Kee somewhat. Looking back, Lee Man Tat believed that Lee Quo-wei had very much facilitated the case, as probably had Richard Charles Lee. He was most grateful to Lee Quo-wei, and until Lee Quo-wei died in 2013, Lee Man Tat often took his children to the top floor of Hang Seng Bank's building to visit and have meals with Lee Quo-wei.

Trade embargo and the temporary suspension of Lee Kum Kee's licence did result in a loss of its business and shocked the whole family. During the event, Lee Shiu Nan and Lee Man Kwong also had to accompany the Director of the Commerce and Industry Ronald Holmes as well as a representative from the United States Consulate to Lau Fau Shan, for the on-site inspection of collecting oysters and making oyster extracts. It was to prove that the raw materials for oyster sauce exported to the United States indeed came from Hong Kong. After the crisis, Lee Man Tat became more vigilant about the sourcing of oyster extracts. He increased the company's sourcing from Lau Fau Shan, and also started to source oyster extracts from Japan. Lee Kum Kee could afford sourcing the Japanese oyster extracts mainly for one reason: the

Japanese, though consume oysters, made no use of their extracts and might simply throw them away.

The crisis had once again proven Lee Man Tat's courage, astuteness and his good skills in networking. He apparently performed much better than his cousins. Over the years, he had experienced lots of conflicts with them over both the means and vision of Lee Kum Kee's development. He might therefore already have envisioned a separation of the Lee family after the crisis. However, he was not financially strong enough to buy shares from the other two. But a new socio-political development of Macau and Hong Kong societies soon created an opportunity to him for the change. The two riots, which almost concurrently happened in Macau and Hong Kong respectively, shocked many in an unexpected way.

The Impact of Two Riots

Macau and Hong Kong were indeed severed from its Chinese motherland in various aspects for a substantial period of time, particularly in the first three decades following the establishment of the People's Republic of China. That said, the two societies remained connected with Mainland China in certain ways and through certain means, and therefore mutually affected at certain historical period.

In short, the 1960s was in general an age of turbulence to Mainland China as the Cultural Revolution started in 1966. The upheavals as well as consequences of the Cultural Revolution had profoundly affected Macau and Hong Kong societies, which directly or indirectly led to the "12-3 Incident"(15) and the following turbulence in Macau in December 1966 and the "1967 Riots"(16) in Hong Kong. The two riots had created a sense of pessimism about the future of Macau and Hong Kong in certain quarters of the societies, and some finally decide to migrate to other countries, particularly to North America and Europe. The

Ronald Holms, the then Director of Commerce and Industry, inspected oyster farming and harvesting in Lau Fau Shan together with representatives from the United States Consulate General.

emigrants from Macau and Hong Kong included relatives of Lee Man Tat and Choi May Ling. However, Lee Man Tat remained confident over the long-term development of Hong Kong given its distinctive role in connecting China and the Chinese communities worldwide. Instead of leaving, he grasped the special opportunity for a breakthrough both by himself and by the family.

Regarding the trend of migration over that period of time, Choi May Ling's second brother had long migrated from Macau to Brazil in the early 1950, while her seventh brother also immigrated to Hawaii around the same time. As for her younger brother Choi Cheong Too, after graduating from the Far East Training School, he moved to work in Hong Kong in 1952. Choi Cheong Too found a good job that matched his professional training in the Hong Kong Aircraft Engineering Company, he then settled down in Hong Kong and got married in 1963. With his wife and children, Choi Cheong Too could have enjoyed a stable and happy life similar to that of Lee Man Tat and Choi May Ling in Hong Kong. He was even promoted to a senior and well-paid position in the company in November 1968, which was a position rarely occupied by the Chinese then. Nevertheless, Choi Cheong Too decided to migrate to Hawaii as well after seeking the consent of his wife, who was a professional nurse in Hong Kong.

But why was Choi Cheong Too so adamant in his decision to emigrate? Choi Cheong Too recalled that on the day of 12-3 Incident (3 December 1966), he happened to be in Macau. He sensed the highly agitated social atmosphere and began to worry about his parents' safety. Since all his brothers were not in Macau, that night he took his parents to Hong Kong for a temporary shelter at his home. In 1967, during the Riots, Choi Cheong Too again encountered something awful. One day, when he was driving off work from Kai Tak Airport back to his home on Man Wan Road, Waterloo Hill, he witnessed the violent incident of Lam Bun, a radio commentator, being burned alive in the car. After experiencing all these incidents, Choi Cheong Too decided to join his elder brother in Hawaii for a more leisurely and quiet life there by running a little store, instead of risking his life in Hong Kong.

For the Lee family, the third generation from Lee Shiu Wing's and Lee Shiu Dang's branches apparently shared a similar pessimistic view and anxiety over the socio-political uncertainties in Macau and Hong Kong. Their solution to uncertainties was to sell all their properties and business in the two places for a better home elsewhere. They began to argue for a bleak prospect of Lee Kum Kee in the region. However, as a matter of fact, Lee Kum Kee's business remained considerably stable during the time, as its core markets in North America were barely affected. Lee Man Tat's cousins thought otherwise, and the disagreement, coupled with their previous discords with Lee Man Tat over the management of Lee Kum Kee, ultimately led to the first family separation.

There were then two camps within the Lee family. On the one hand, it was offspring of Lee Shiu Wing and Lee Shiu Dang who wished to sell the business and their properties and migrate to other countries. On the other hand, it was Lee Man Tat, with the full support of his father Lee Shiu Nan, who wished to continue the almost century-old Lee Kum Kee. The old-fashioned Lee Shiu Nan would never agree to the family separation, as he cherished his father's business and reputation very much. He once appealed to Lee Shiu Dang's widow for help, but to no avail. Lee Man Tat eventually acquired from the other camp their shares of Lee Kum Kee, and marched to the next stage of his career, even though it was a hard time at the beginning of the division. The transaction was expensive and beyond his means, hence he had to find a way out and boost the business as quickly as possible.

Conclusion

Life cannot be a bed of all roses; only those who are gallant and persistent succeed to make their way to the promised land. Lee Man Tat was weak at birth and worrying to his parents, but such was the vitality in his life blood that he subsequently grew up strong and healthy. When Lee Man Tat joined Lee Kum Kee in 1955 and moved from Macau to Hong Kong, he was expecting an exciting new chapter in his life. But he later found out that it was an experience far below his expectation. He nevertheless contributed to the settlement of a serious crisis that almost engulfed Lee Kum Kee, and had tried his best to reform the family business within the limit. His ability and talent in business again impressed his parents as well as his cousins.

Undoubtedly, the ambitious and dynamic Lee Man Tat never stopped exploring other business opportunities throughout the 1950s and 1960s. It was also a period that he gradually built up his personal network, which greatly facilitated his business later. He loves to socialise with others and always manages the relations with a sincere heart. The power of his network would soon be put onto another test.

1   The closest family members of Lee Man Tat, including wife Choi May Ling, and five children Eddy, Elizabeth, David, Charlie and Sammy, all vividly described his almost unlimited stamina and passion in doing business in their interviews with the authors. Lee Man Tat's closest relatives, friends, business partners and subordinates, for instance Tang Fook Chuen, Choi Cheong Too, Chang Chin Vai, Chow Sing Lam, Vincent Ko, Cliff Ho, Pao Pui Li and Loi Kam Sheung, also shared with us in the interviews how energetic and efficient Lee Man Tat was in running business.

2   At that time, Chang Chin Vai was left as the only staff minding the old Lee Kum Kee store in Macau. It was selling only oyster sauce and shrimp paste, and the turnover was pretty low. Its meagre revenue was not even sufficient to cover Chang Chin Vai's monthly salary, thus additional financial support was needed from Hong Kong.

3   The International Funeral Parlour was opened in February 1951 at No. 41-51 of Lockhart Road in Wan Chai. It was then a single-storey building, with its ground floor as the hall and basement as the morgue. However, a strong demand for more spaces prompted the construction of a five-storey new building at No. 16 Hoi Tai Street in North Point in 1960, and the funeral parlour was formally relocated in July 1962. Its Wan Chai site was converted into East Town Theatre, but the theatre was allegedly haunted and lack of audience. It was finally re-constructed as East Town Building.

4   Lee Suen Funeral Parlour was located at No. 10 Morrison Hill Road, Hong Kong Funeral Home at No. 216 Wan Chai Road, Morrison Funeral Parlour at No. 4-8 Morrison Hill Road and No. 559-561 Nathan Road, and Fok Sau Funeral Parlour at No. 189 Queen's Road East.

5   Ho Tim was one of the founders of Hang Seng Bank of Hong Kong. Together with Lam Bing Yim, Ho Sin Hang, Leung Chik Wai and Sheng Tsun Lin, the five co-founded "Hang Seng Ngan Ho" in the 1930s, which was the predecessor of today's Hang Seng Bank.

6   Liang Yuen Cheong, better known as Liang Cheong, was a wealthy businessman in Hong Kong and Macau during the 1950s and 1960s. He earned his first fortune through foreign trade, particularly concerning the trading of gold. He later served as a director of the Hang Seng Bank of Hong Kong.

7   "Radio Rediffusion" was the first privately-run radio broadcasting station in Hong Kong which kicked off its broadcasting service on 22 March 1949. Its installation fee at

the time was as high as HK$ 25; moreover, the audience would have to pay HK$ 9 per month. It was therefore not an affordable entertainment to every household. Radio Rediffusion offered only two radio channels, namely "Channel Silver" which was mainly in Cantonese, and "Channel Blue" which was mainly in English. The broadcasting service offered 24 hours of programmes around the clock.

8   The old building was put on hold for many years without any development. But it was finally re-built as today's "Hollywood Terrace."

9   "Salted shrimp paste" was one of the few products of Lee Kum Kee at the time. It is a kind of coarse shrimp paste, in which tiny eyeballs of the shrimps could still be seen. In contrast, shrimp paste is finer.

10   Quote from her siblings' sharing on the occasion of the "60th wedding anniversary cum 85th birthday banquet of Lee Man Tat and Choi May Ling" on 23 December 2014.

11   Due to distraction caused by the May-Fourth Movement, Liu Tai-shi missed the entrance examination of Peking University in 1919. He didn't want to give up, therefore decided to seek audience with Tsai Yuan-pei, the then President of Peking University with his high school's composition and poetry collections, hoping for a second chance of admission. After reading his composition pieces and poetry, Tsai Yuan-pei dictated a title to him, and asked him to write something accordingly in his office. Liu Tai-shi quickly finished the task. Tsai Yuan-pei was impressed by his article, thus immediately decided to recruit him into the Arts Foundation Courses.

12   Liang Zhangju was a scholar as well as an imperial official from the Qing Dynasty.

13   Quoted from Lee Man Tat at the occasion of the "60th wedding anniversary cum 85th birthday banquet of Lee Man Tat and Choi May Ling" on 23 December 2014.

14   Lee Quo-wei was a famous Hong Kong banker. He was then the General Manager of Hang Seng Bank, a social elite that the British colonial government was always keen to co-opt into its establishment. He was appointed as an unofficial member of the Legislative and Executive Councils in 1968 and 1976, but withdrew from the two Councils in 1978. Lee Quo-wei was Chairman of Hang Seng Bank between 1983 and 1997, and re-appointed as an unofficial member of the Executive Council between 1983 and 1988. Due to his prominent standing, Lee Quo-wei also served as Chairman of the Council of the

Chinese University of Hong Kong between 1982 and 1997.

15   The "12-3 Incident" of Macau had its root in a November 1966 conflict between Macau police and the local people. In November 1966, the Taipa residents of Macau were stopped by the municipal authorities while building their own public primary schools, because they did not apply for a permit to set bamboo scaffolding in advance. The locals were upset by such restriction and refused to back down. On November 15, the Portuguese Macau government sent out military police to dispel people and wounded 34. On December 3, when representatives from various sectors petitioned to the Governor of Macau, they were again beaten by the military police. It aroused public anger, and riots subsequent erupted in the city. The crackdown again caused casualties. After several rounds of serious clashes between the two parties, the Portuguese Macau government was forced to accept all requests from the local Chinese community unconditionally on 29 January 1967.

16   The "1967 Riots" first erupted on 6 May 1967 and ended in December of the same year. It could be seen as a local uprising against the British colonial government under the influence of the extreme ideology of the Chinese Cultural Revolution. It was initially a workers' movement, but gradually escalated to anti-government demonstrations, and finally turned into terrorist actions including bomb attacks, killing and injuring many people.

Lee Man Tat (2nd right) and friends posed for a photo in front of the Wong Chuk Hang plant in 1976.

RISE IN CRISES

I was actually quite depressed for a period of nearly two decades between 1954 and 1972 because other family members were simply reluctant to make Lee Kum Kee stronger and better. All the views raised by them were very conservative. My cousin Lee Man Tip was previously responsible for the company's sales, unfortunately he died prematurely in 1964. The business of Lee Kum Kee began to decline after that, and its sales went downhill all the way. If I didn't take over to reform, the company would eventually be eliminated.

Lee Man Tat

INTRODUCTION

After devoting himself full-time to Lee Kum Kee, Lee Man Tat gradually learnt from experience the competitive edge and intrinsic limitations of a family business. He also had a deeper understanding of the nature of social relations while building up his personal network and the secret to breakthrough in business ventures. "Si Li Ji Ren" (Considering Others' Interests) was quite a quintessential summary of his personal reflections in the field. His steadfastness and sagacity in times of adversity was apparent during the crisis of oyster extracts in 1965 as was the quality of his leadership which stood out glaringly amongst the third generation of the Lee family.

While facing difficulties and handling crisis after crisis, a notable attribute of Lee Man Tat came to light – that is, his unique ability to see the big picture from the small scene, or to put it differently, "seeing the universe from a grain of sand". In the following chapters, we can see that Lee Man Tat has always been able to grasp opportunities from crises while arduously tackling them, and subsequently led the family and its enterprise to a higher plateau. The Lee family ran into a greater crisis in the late 1960s, but unlike the saga of oyster extracts in 1965, it was a crisis from within. The incident was critical in the sense that if failed to be handled judiciously, Lee Kum Kee, an 80-year old family business which two generations of his forefathers had put their heart and soul into building up, would simply disintegrate and disappear beyond the scene. Visualising the intricacy of the problems, Lee Man Tat made the right move to acquire the sole ownership of Lee Kum Kee. The decision not only enabled him to resolve the crisis but also to implement his reform ideas and drive the business of Lee Kum Kee to new heights.

The First Family Separation

Crises may turn into opportunities if handled sensibly in a right historical context. In retrospect, the separation of the three branches of the Lee family in 1972 was in fact the most significant turning point in the career of Lee Man Tat. Before that, he was passionate to reinvigorate Lee Kum Kee but was dragged down by his cousins. It was a big family and as an offspring of Lee Shiu Nan, the third and youngest son of Lee Kum Sheung, he had barely any hopes to assume leadership. Opinions and views were also diverse within the family, and subtle discords could always be found among branches of the family of the second generation. After 1972, Lee Man Tat took the helm of Lee Kum Kee and, under a unified leadership, he could follow his own judgment in implementing the reform blueprint. It was a blueprint that he had long thought of, which in fact led to the incarnation of "Lee Man Tat's Lee Kum Kee."

Lee Kum Kee was at the time a typical Chinese family business run in a highly traditional Chinese way. Understandably, Lee Shiu Nan was set in the most disadvantaged position as the youngest son in the second generation. His eldest brother, Lee Shiu Wing, died prematurely during the World War II. But Lee Shiu Wing had four grown-up sons, and amongst them, Lee Man Kwong, Lee Man Tip and Lee Man On had long participated in Lee Kum Kee's business. The licensee of Lee Kum Kee, as we could see during the crisis of oyster extracts in 1965, was Lee Shiu Wing's eldest son Lee Man Kwong.

Lee Shiu Dang, the second son in the second generation, died in the early 1954. He had two sons, namely Lee Man Shiu and Lee Chee Ping, but the two did not actively involve in Lee Kum Kee's business. Lee Shiu Nan himself had two sons, Lee Man Tat is the elder one and Lee Man Lok the second (ranked fourth amongst his siblings). Lee Man Tat was more than ten years old than Lee Man Lok, thus it was Lee Man Tat alone who participated in Lee Kum Kee's business since the-mid 1950s.

Lee Shiu Nan could in fact be regarded as the pillar of Lee Kum Kee. In the early days of Lee Kum Kee, he assisted his father Lee Kum Sheung to improve the formulation of oyster sauce, and when Lee Kum Sheung passed away, he took charge of the workshop and production, followed strictly and meticulously his father's principles on quality control, and kept improving every detail of the production. This was the foundation that sustained Lee Kum Kee's longstanding reputation. Lee Shiu Nan was by nature a man who tended to avoid strife and conflicts. He therefore discouraged his eldest son to get involved in Lee Kum Kee's business after the World War II, even though young Lee Man Tat was more than eager to have a try. Instead, he urged Lee Man Tat to go to Guangzhou to make his mark there.

Regarding his second son Lee Man Lok, Lee Shiu Nan did not arrange for him to join Lee Kum Kee either. Lee Man Lok graduated from St. Stephen's College in Hong Kong in the late 1950s, he was introduced by Lee Man Tat to Hang Seng Bank for a stable job in a different career. After Lee Shiu Dang had passed away in 1954, Lee Shiu Nan as the only son from the second generation enjoyed a higher and more respected status vis-à-vis his nephews, and took charge of the company's accounts books. Having said that, Lee Shiu Nan still did not envision a dominance of his sons in Lee Kum Kee in the future, as the idea to him was simply unrealistic. He retired later as he advanced in age, leaving Lee Kum Kee to the third generation.

Notwithstanding his frustration for not being able to realise his dreams, Lee Man Tat's passion for the Lee Kum Kee family business never diminished. He had mastered every step of the oyster sauce production and techniques concerned since young. As a gourmet, he had great taste buds as well, which enabled him to distinguish good oyster sauce and shrimp pastes from the bad. But what was more important was his optimism towards the sauce and condiment industry. As he put it, although it was not difficult for outsiders to cross over the technical barrier and join the industry, it was never easy to excel in the field. "The rich generally don't want to get involved in this business because they have to deal with the multifarious problems of sourcing raw materials, the production, as well as the sales." In contrast, Lee Kum Kee was already well-established and leading in the industry, and that implied an even brighter future for it.

In particular, Lee Man Tat was astutely aware that sauces and condiments were necessities of daily lives: every household would consume them. His conviction was simple: if we made it right, the market could be infinitely large. There was therefore no reason whatever to give up or wrap up Lee Kum Kee. Based on such optimistic perspective, he always attempted to aggressively explore new markets for Lee Kum Kee in the 1950s and 1960s, but none of his cousins shared his passion. In the interview, Lee Man Tat sighed that:

> *I was actually quite depressed for a period of nearly two decades between 1954 and 1972 because other family members were simply reluctant to make Lee Kum Kee stronger and better. All the views raised by them were very conservative. My cousin Lee Man Tip was previously responsible for the company's sales, unfortunately he died prematurely in 1964. The business of Lee Kum Kee began to decline after that, and its sales went downhill all the way. If I didn't take over to reform, the company would eventually be eliminated.*

It was quite likely that Lee Man Tat had come up with the idea of buying shares from all the other family members and to take charge of Lee Kum Kee solely, but his capital on hand was not enough for such takeover. Lee Man Tat had earned his first bucket of gold in the 1950s in Macau from his trading and manufacturing business. He invested extensively in various fields such as the funeral industry, real estate and catering in the 1960s, and effectively accumulated much capital from these investments. But to buy shares of Lee Kum Kee from all his cousins would require a lot of money: while Lee Kum Kee was not a big company, it was neither a small one. In short, Lee Man Tat was still financially incapable to make such proposal by the mid-1960s.

In Sole Charge of Lee Kum Kee

The unexpected turbulence which hard hit Hong Kong and Macau and created shock waves to the people in the late 1960s changed the balance within the Lee family. In fact, the riots that erupted in both societies did not last long. To Lee Man Tat, it was a rare historic moment to seize in concluding a deal for the takeover of the business from other shareholders and run it his own way. He grasped the opportunity without any hesitation.

The outbreak of the decade long Cultural Revolution in Mainland China in May 1966 quickly impacted the neighbouring Macau and Hong Kong communities, particularly amongst the local youngsters. Consequently, the "12-3 Incident" erupted in Macau in December 1966 due to untoward handling of the Portuguese Macau administration on a police-civilian conflict. A series of bloody conflicts continued after the Incident, which gravely affected governance of the Portuguese Macau administration. Finally, the civil unrest ended on 29 January 1967 when the Macau Government unconditionally accepted all demands made by the local Chinese community.

Macau and Hong Kong had long been neglected under the colonial rules of the Portuguese and British respectively. Community conflicts in the two societies, coupled with the radical and aggressive atmosphere of class struggle fueled by the Cultural Revolution, prompted agitation. The "Kowloon Disturbances" broke out in April 1966 following a series of protests against fare increase of the Star Ferry were regarded as a local incident without apparent external influence or intervention, but they definitely boded ill for the Hong Kong society. When the flare of Cultural Revolution began to burn in Mainland China, Liao Chengzhi, the experienced and steady man in charge of Hong Kong and Macau affairs in the State Council, lost power. Riots subsequently erupted in Hong Kong in May 1967.

Although the riots ended in December 1967 and the British colonial government survived, Hong Kong's future was clouded by uncertainty. Lee Man Tat's cousins, like himself, had experienced the severe negative onslaughts of the 1949 regime change in China. And now, in the late 1960s, with their stronger financial capability to go far away to the United States and Canada, they began to hatch the idea of selling out ancestral assets for migration. Lee Man Tat however had a different view on this.

First of all, he opined that even though Hong Kong's economy was relative volatile in the 1960s and 1970s, their overseas markets in North America were then the major clientele of Lee Kum Kee, thus turn-over in the Hong Kong market would not significantly affect its overall company performance. Moreover, as his strategy of launching a new economical brand of oyster sauce aimed primarily at the United States market, he reckoned that they could bear the brunt of social unrest in Hong Kong. And thirdly, he was still confident of the stability of the British colonial regime at the time. He judged that the Chinese Communist regime might not want to take over Hong Kong even during the turbulent time of Cultural Revolution, or to see Hong Kong trapped in turmoil. Social unrest in Hong Kong, in that sense, should fade away soon.

Based on the above observation and consideration regarding the socio-political changes in Hong Kong and Macau, Lee Man Tat surprised many in his aggressive investments around the time of the 1967 disturbances. For instance, he invested a huge sum to purchase one of five office blocks of the General Commercial Building, located at No. 156-164 Des Voeux Road Central, Central. It was a 15-storey building, and Lee Man Tat acquired the building for rental income as a long-term investment property instead of short-term speculation. The investment action clearly revealed his strong confidence in the future of Hong Kong.

According to Lee Man Tat, when the family members began to negotiate details of the separation, members from Lee Shiu Wing' and Lee Shiu Dang's branches joined hands to pressure Lee Shiu Nan and his sons in selling them their shares. Lee Man Tat adamantly rejected the proposal with a blunt reply: "This is the business that I am running, I will never sell it to you." The response again showed that oyster sauce business was his true love, period. After much deliberation, they eventually reached the consensus that Lee Kum Kee Company was worth HK$ 6.9 million. Based on such evaluation, each of the three branches should have its share of HK$ 2.3 million. If the other two branches united as a camp to buy shares from Lee Shiu Nan's branch, each of them would only need to pay HK$ 1.15 million. In contrast, if Lee Shiu Nan wanted to buy their shares, he would need to make ready a huge sum of HK$ 4.6 million. "But how could we have so much money?" Lee Man Tat lamented while recalling the family separation. As a result, the negotiation came to a deadlock. During the stalemate, proposals of "splitting Lee Kum Kee into two or three companies with their respective brand names" were even once contemplated. It was a critical moment to Lee Kum Kee, as it might very likely fall into the abyss of division and even closing down eventually.

At the beginning of the separation negotiations, Lee Shiu Nan's view obviously differed from his eldest son. Lee Man Tat noted that although the three branches often disagreed in views, his father remained very traditional in his perspective of family. Lee Shiu Nan was adamant that

Social turbulences in 1967 prompted many people to leave Hong Kong. Lee Man Tat, on the contrary, decided to take what others had discarded and worked even harder to pursue his ambitious goal. (Source of photo: Ko Tim Keung)

"Lee Kum Kee could only sustain its prosperity with a united family. A divided family will be the crack of doom to Lee Kum Kee. Its legacy from generation to generation will then be over." He thus refused to consider any option of family separation, with a strong fear that the legacy of his dear father Lee Kum Sheung, who laid the foundation of Lee Kum Kee single-handedly, would be ruined in the hands of the third generation.

Chang Chin Vai today still vividly remembers the scene of Lee Shiu Nan's visit to his sister-in-law, the widow of Lee Shiu Dang. Lee Shiu Nan went to her home together with him and another staff from the store in Macau. He prompted her to take charge of the situation to avoid family separation, with a stern warning that family separation would weaken the business of Lee Kum Kee. To his surprise, his sister-in-law indifferently replied, "I can't hold these people anymore: just let it be." The unexpected response from Lee Shiu Dang's widow dawned on Lee Shiu Nan that they might have made up their mind for the separation. He left her home with a heavy heart and tears welled up in his eyes, deeply disappointed.

After half a year of impasse, the two camps finally reached a deal. After all, the other two branches no longer had the patience to run the medium-sized and "future uncertain" Lee Kum Kee and were eager to leave Hong Kong after the 1967 disturbances. They accepted the terms of Lee Man Tat in paying them the amount of HK$ 4.6 million in installments. The first family separation of Lee Kum Kee was subsequently settled. On 14 July 1971, members of the three branches signed the documents in the office of Deacons law firm, witnessed by their lawyers. Notice of the "dissolution of partnership" of Lee Kum Kee was then published in major Chinese and English newspapers in Hong Kong as follows:

LEE KUM (OR KAM) KEE (李錦記)

NOTICE OF DISSOLUTION OF PARTNERSHIP

WHEREAS:
The abovenamed firm has hitherto been carrying on business at No. 207, Queen's Road Central, ground floor, Hong Kong by LEE SHIU NAM (李兆南), LEE MAN TAT (李文達) and LEE MAN LOK (李文 樂) (hereinafter referred as "the Continuing Partners") and LEE MAN KWONG (李文光), LEE MAN ON (李文安) and the administrators of LEE MAN TIP, DECEASED, (MADAM LEE CHING MING (李 正明) and MR. LEE MAN ON (李文安), LEE MAN SHIU (李民 紹) and LEE CHEE PING (李治平) (hereinafter referred to as "the Retiring Partners") in partnership at will for the manufacturing and wholesale of oyster sauce, shrimp paste and other food products.

AND WHEREAS:
The said parties have at the date hereof executed a Deed of Dissolution of the said partnership at the Offices of Messrs. Deacons, solicitors, whereby the Retiring Partners assign and release their shares and interest of and in the said business of the said partnership to the Continuing Partners to the intent that the Continuing Partners shall henceforth be the only continuing partners of the said Lee Kum (or Kam) Kee.

NOTICE IS HEREBY GIVEN that the dissolution of the said partnership takes effect as from the 14th July 1971, and that the business of the said Lee Kum (or Kam) Kee shall henceforth be carried on by the Continuing Partners who shall be responsible for all debts and liabilities due by the said Lee Kum (or Kam) Kee prior to as well as after the date of the dissolution AND FURTHER that the Retiring Partners shall hereafter not be responsible for all the debts and incumbrances in connection with the said Lee Kum (or Kam) Kee.
Dated this 14th day of the July 1971.

Messrs. Deacons,
Solicitors for the Continuing Partners and the Retiring Partners,
601 Union House, HONG KONG.

To this day, Lee Man Tat clearly remembers that since Lee Kum Kee was not a limited company at that time, all family members involved in the deal had to be present and sign the documents. "Hence, in the office of Deacons law firm in Union House, Hong Kong, we followed the explanation and instruction of the lawyers and signed one document after another. It was a long drawn process running from 4 pm to around 11 pm." In the agreement of the dissolution of partnership, the other two branches signed out of Lee Kum Kee, whereas Lee Man Tat signed on behalf of the third party to take up their released shares. Lee Kum Kee, from then on, has embarked on the "Lee Man Tat Era." Under the strong leadership of Lee Man Tat, the then medium-sized Lee Kum Kee has morphed into a multinational corporation with worldwide presence of its products in just around three decades.

Dispense with Kam Shan Chuang

The two policies implemented immediately after Lee Man Tat took the helm of Lee Kum Kee in 1972 were inter-related: he wanted to get rid of Kam Shan Chuang, while at the same time, launch the economical oyster sauce. They were closely related to each other because, in order to operate free from Kam Shan Chuang, he needed the economical oyster sauce to greatly expand the market of oyster sauce, and a bigger market entailed a more extensive and effective network of sales. If he failed to dissociate from Kam Shan Chuang, his sales in the market would still be very much restricted.

Lee Man Tat had learnt from bitter experience the devastating strength of the monopoly of Kam Shan Chuang as early as the 1950s and 1960s. His most important task then was not direct liaison with Lee Kum Kee's

李錦記承退股啟事

李錦記为向由李兆南李文逵李文樂(下文統稱「承股人」)及李文光,李文安,李文鑾之承辦人(李正明及李文安退股人」)合股在香港大道中二○七號地下經營蠔油蝦醬等食物生意,現雙方已於卽日在的近律師行簽約退股,承股人退出該李錦記將其名下之股份權利全部讓與承股人今後由承股人全部經營該李錦記之生意並負責李錦記以前現在及卽後之債務,此後有關李錦記之華洋債項與退股人無涉,此啟。

一九七一年七月十四日

香港於仁大廈六○一室

退股人之代表律師的近律師行

"Notice of Dissolution of Partnership": this event marked a turning point of Lee Kum Kee.

distributors, clients or consumers in North America, but frequent visits to Kam Shan Chuang traders and socialise with them because they controlled the sales networks among overseas Chinese communities in North America. Lee Man Tat was particularly dissatisfied with Kam Shan Chuang earning commissions from both sides. Their exorbitant profits seriously exploited revenue of suppliers and unduly raised the price for the consumers as well. Moreover, in order to appease them, suppliers were required to present lavish gifts to Kam Shan Chuang over Chinese New Year and other festivals to increase their indirect costs.

Eddy Lee and David Lee still remember that during their childhood, on the first day of every Chinese New Year, their father Lee Man Tat would first pay visits to bosses of Kam Shan Chuang to present them gifts before they paid respects to their grandfather Lee Shiu Nan and other relatives. Lee Man Tat had to bow to Kam Shan Chuang because they held the lifeline on Lee Kum Kee's sales, even though it was Lee Kum Kee's oyster sauce that contributed to a significant profit of Kam Shan Chuang. The sales operation of Lee Kum Kee in the United States was thus unfairly constrained by the Kam Shan Chuang.

Kam Shan Chuang was a unique historical product of Hong Kong. It reflected the unique intermediary role played by Hong Kong in connecting China and the world since it opened for trade in the mid-19th century. Before the World War II, apart from Kam Shan Chuang, Hong Kong also had a bunch of other trading agents with different regional focuses, including trading agents to Sydney, Peru, Japan, the Philippines, Vietnam, Penang, Java Island and Surabaya.

Kam Shan Chuang specialised in serving overseas Chinese communities in San Francisco, Los Angeles and New York in the United States, as well as Vancouver in Canada. These were all North American cities with populated Chinese migrants. The commodities that Kam Shan Chuang purchased and sent from China on behalf of their North

American clients were highly diverse, entirely dependent on the needs of their clients. Nevertheless, these commodities could be broadly categorised as Chinese food (e.g. preserved duck, salted duck egg and various condiments), Chinese medicine, and special local products from the Pearl River Delta.

Kam Shan Chuang usually also took charge of the accommodation of Southern Chinese labours in Hong Kong while they were on their way to North America. Upon their arrival in North America, Kam Shan Chuang would offer them services of remittance and to send letters home (Zheng, 2016: 50-52).

After the first Opium War in the 1840s, the Qing government of China was forced to open up five of its ports on its South-eastern coast for foreign trade, which triggered waves of emigration of labours from the countryside of South China. The Chinese labours, at that time, aimed predominantly for the three destinations of Kam Shan (referred in a broader sense to the United States), Nam Yong (Southeast Asia) and San Kam Shan (referred in a broader sense to Australia). Thus decades later, by the early 1880s, Kam Shan Chuang such as Kwong Yuk Fong, Kwong Wing Sang and Thai Hing Long appeared in Hong Kong. By the 1890s, Kam Shan Chuang had reached its height with a number of more than a hundred, as they spread across an area stretching from the waterfront of Sheung Wan, including Queen's Road Central, Queen's Road West, Des Voeux Road Central, Des Voeux Road West, Bonham Strand, Wing Lok Street and Connaught Road. (Zheng, 2016: 50-51) They once out-numbered rice shops and banks which were in fact more critical to the daily lives of the people, and became a unique commercial phenomenon.

By the late 1930s, around 150 Kam Shan Chuang were still operating in Hong Kong. However, with the cloud of war hanging over at that time, their business had become more difficult. When the Japanese occupied Hong Kong from 1942-1945, Kam Shan Chuang in Hong Kong were unable to trade with the United States; most of them therefore shifted their operations to Macau and even Guangzhouwan, which is today's Zhanjiang City (Zheng, 2016: 52). By 1952, business of Kam Shan Chuang was again hit by the escalating U.S. led-trade embargo on China following the outbreak of Korean War. Since then, the industry had begun to decline, although its demise was gradual.

Amongst the Kam Shan Chuang, Lee Kum Kee had a closer partnership with Wang Yue Long and Jaxon.(2) According to Loi Kam Sheung, whose father and grandfather were both running the Jaxon, Kam Shan Chuang was actually the core of Hong Kong's entrepot trade and was highly important to its economic development. If we take Jaxon as an example, the review of its trading commodities shows that for all its exports to San Francisco's Chinese community, condiment made up 80%, and amongst those 80%, oyster sauce was the main item. Lee Kum Kee was once associated with more than twenty Kam Shan Chuang in the United States, and in comparison, the number of associated Kam Shan Chuang in New York on the East Coast was higher than that in San Francisco on the West Coast.

Lee Man Tat realised that in order to side-step Kam Shan Chuang, Lee Kum Kee must build its own sales and distribution networks, and make inroads directly into the North American market. Lee Man Tat had raised such idea to his cousins as early as the 1960s, but none of them dared to risk souring their relationship and, more importantly, they were scared to face any mishap during the period of transition. Kam Shan Chuang indeed provided effective intermediary services, but they were not exclusive agents for Lee Kum Kee, thus when listing their trading commodities, they would not think from the perspective of Lee Kum Kee's interest, not to mention managing Lee Kum Kee's customer relations in North America on its behalf and paying attention to its clients' feedback. Moreover, it was very difficult for Lee Kum Kee

Accounts books that recorded transactions with Kam Shan Chong
are now part of the treasured historical archives of Lee Kum Kee.

to grasp updated information of its market sales, customers' preference and the positioning of products via Kam Shan Chuang, and to plan its production more accurately.

Although Lee Man Tat had made up his mind to cease relying on Kam Shan Chuang, he realised the risk at stake and did not want to act recklessly. His first move, after much deliberation, was to divert the attention of Kam Shan Chuang by introducing a new brand of economical oyster sauce called "Panda." This move also served a dual purpose for Lee Kum Kee, after so many years of selling only one kind of high-end quality oyster sauce product, to launch an economical "Panda" brand.

Promoting Panda Brand Oyster Sauce

Economical oyster sauce was not a new idea that Lee Man Tat had just come up with. His strategic rationale was simple: a premium product with excellent quality was good but had its limitation in dominating the mass consumer market, and hence in increasing sales volume. He would like to present both the premium and the economical oyster sauces to the market to cater to different customer groups. However, Lee Man Tat himself insisted, "a cheaper oyster sauce brand does not mean that we simply add water to dilute the sauce, or follow others' practice of adopting inferior raw materials." The umami or savoury taste of oyster sauce, to him, must never be compromised.

Lee Man Tat needed to work out a revised formula for the economical oyster sauce. But for a sauce company such as Lee Kum Kee, formulae of its products were extremely sensitive information, so he simply could not entrust the task to any outsider. Lee Man Tat then decided to seek help from a retired brother-in-law, Choi Cheung Yiu, who was the fourth brother of Choi May Ling. Choi Cheung Yiu graduated from the Chemistry Department of University of Nanking. He accepted the invitation and subsequently applied his professional knowledge and expertise to work out a formula for Lee Kum Kee's economical oyster sauce. As for the brand name of "Panda", indeed it related to a contemporary event with monumental significance, but it was a 'snap of the moment' decision without much depth of thought.

It was in February 1972 — the same year that Lee Man Tat assumed leadership of Lee Kum Kee — that the U.S. President Richard Milhous Nixon made his ice-breaking trip to China during the period of Cold War and, as a result, the two countries moved towards normalisation of relations. The visit was a momentous turning point in the tripartite relationship amongst the United States, the Soviet Union and China during the Cold War, and thus attracted worldwide attention. On the last night of Nixon's visit to China, Zhou Enlai, on behalf of the Chinese government, announced the gift of a pair of giant pandas to the United States, which became the first case of China's "panda diplomacy." The two pandas enjoyed extensive attention of the international community, particularly the Americans.

When the two pandas finally settled down in the National Zoo in Washington, their lively and lovely gestures soon aroused the curiosity and passion of the American people. It was around the same time that Lee Man Tat began to focus his mind on marketing his economical oyster sauce to the United Sates market. Thus by coincidence rather than designed, he seized the moment to adopt "Panda" as the brand name, which he thought might be a trendy brand name to the Americans.

Lee Man Tat quickly set for detailed preparation in 1972 and when he thought everything was ready, he launched the Panda Brand oyster sauce first to the North America market. He was pragmatic enough to market concurrently through two different distribution channels for the sales of the new products, namely, the old network of Kam

Shan Chuang and the new network established by Lee Kum Kee itself. For the latter, he flew to several major cities in the United States to personally visit grocery store and restaurant owners in the local Asian community. His intention was to first show them respect and secondly, directly promote the economical oyster sauce to them. Moreover, he also tried to invite them to become local sales agents for Lee Kum Kee, thus facilitate the establishment of its own sales network. However, immediately after the launch of panda brand oyster sauce, it did not become a hot item as he had hoped, and indeed dealt a somewhat heavy blow to Lee Man Tat. After the first round of selling, the old network of Kam Shan Chuang still mainly asked for Premium Oyster Sauce and shrimp paste instead of the new product. Neither did Lee Man Tat manage to build up his own sales network as planned.

Lee Man Tat refused to accept that it was a fiasco. He remained confident in his strategy of economical oyster sauce as he knew that the Panda Brand, even though a cheaper kind of oyster sauce, was good in quality. During his time in the United States, he once spent a few days wholly in Chinatown supermarkets in Los Angeles and New York for answers. He soon convinced himself that he had not made the wrong decision because the market in general was receptive to the economical oyster sauce, and what was needed was improvement to its promotion details and product packaging. For example, the original colour background of Panda Brand was yellow which was not bright and attractive enough in the dazzling range of goods. Lee Man Tat quickly sent a fax to Hong Kong and requested the staff to change its colour background from yellow to red.

On the other hand, he approached regional distributors as well as owners of grocery stores and supermarkets in Asian communities, and offered them something they could hardly resist: selling with "credit consignment." Lee Kum Kee's products, including the Panda Brand oyster sauce, would now be sent to them for sale before payment. They

The logo of the Panda Brand Oyster Sauce (above) and the old logo of the Premium Oyster Sauce (below). Lee Man Tat was involved in the design of both logos.

could delay payment until the goods were sold. Such tactic of sales promotion instantly boosted the enthusiasm of both its distributors and shopkeepers in selling Lee Kum Kee's products. It was a typically daring and risky move of Lee Man Tat that he always loves to share with others: "If they asked for one container of Lee Kum Kee's products, I would send them two. If they asked for five containers, I would send then ten." The brave move and other marketing adjustments finally boosted the sales of Panda Brand oyster sauce. Lee Man Tat did not hide his pride and satisfaction when he said, "These distributors soon came to me with their money to settle their bills. They said that the goods were all sold out, and they needed more: they didn't expect such a good sale." (Wu, 2013: 82)

Eddy Lee, Charlie Lee and Elizabeth Mok all pointed out that their father repeatedly adopted such tactic of selling with "credit consignment" in the time that followed. It was virtually tailor-made for the rising grocery storekeepers in the Asian communities of the United States at the time. While the sales model was slightly adjusted later, it still worked well to generate satisfactory results. Note that most of the small storekeepers in Asian communities were short of cash in the 1970s, which could affect their ability to prepay goods for sale. Lee Man Tat's way of "sell first, pay later" was indeed very appealing to them and encouraged them to sell more Lee Kum Kee products.

It is worth pointing out that the 1970s also witnessed a new wave of Chinese emigration from Hong Kong, Taiwan and Southeast Asia to North America. Unlike the earlier Chinese workers from South China, the new arrivals were mostly middle-class or highly educated, with a certain degree of economic capability, thus revitalizing the traditional Chinese communities in North America. A booming Chinese catering industry in North America was a result of this new wave of Chinese migration. Chinese restaurants there were then dominated by Cantonese cuisine, and both oyster sauce and seafood sauce were widely used in these Chinese restaurants, thus escalated the sales of Lee Kum Kee's oyster sauce.(3)

Ever since Lee Man Tat liaised directly with distributors and clients in North America, he paid even more attention to Lee Kum Kee's credibility and branding. A distributor in the United States once complained about the flavour of a batch of Lee Kum Kee's oyster sauce. Although Lee Man Tat knew it was not an issue of quality, he recalled the goods immediately, and sent out a different batch to him. The transaction cost Lee Man Tat dearly, but he believed it was definitely worthwhile for the reputation of Lee Kum Kee (Wu, 2013). Little wonder that with Lee Man Tat's insistence that branding is the lifeblood of Lee Kum Kee, their products enjoy a lion share of the Chinese condiments market in North America up to this day: its starting point may well be here.

Time of the Fourth Generation in North America

The five children of Lee Man Tat and Choi May Ling, i.e. the fourth generation of Lee Kum Kee, all completed their primary education in Hong Kong. But by and large they completed their secondary and tertiary education in Canada and the United States. In order to take better care of them and see through their adolescent years, Choi May Ling also spent much time with them in North America in the 1970s. Indeed, they became a pillar of strength to exploit the North America market when Lee Kum Kee went through accelerated growth in the 1980s. As Lee Man Tat recognised, the phenomenal growth in the market share of Lee Kum Kee at that time was largely due to the strong support of his five children and, in many ways, their success was highly related to their education and living experience in North America.

When Eddy Lee completed his junior high school in Hong Kong in 1972, he was the first in the family to study abroad. He went to a boarding school in Victoria, Canada. Choi May Ling preferred Canada over other destinations mainly because she had a nephew who worked as a doctor there and could be around to take care of Eddy. Sharing his short stay in the school with them, Eddy Lee recalled, "Three months after school term started, when my parents came to visit me, they heard the boisterous sound of mahjong before stepping in the dormitory. My father was furious. I explained to him that it was none of my business but other students playing. However, my father was dissatisfied with the school's environment, and asked me to transfer to another school by the end of the semester."

Choi May Ling's nephew therefore recommended another boarding school to them, which was located around 100 miles from Toronto. Eddy Lee went to the school but instead of staying in its dormitory, he was arranged by Choi Mai Ling to live in a small town near the school. Eddy Lee continued, "My father was also not happy with the environment of the small town which he thought was too small and isolated: it was so peaceful that people become relaxed and took things easy, losing their fighting spirit. He therefore prompted my second transfer."

In 1973, Elizabeth Mok and David Lee also went abroad; they attended different schools due to different interests. These schools were actually selected and recommended by their private tutor. Elizabeth Mok spent a year at a boarding school in Hawaii, where two of her uncles resided. David Lee went to the remote province of New Brunswick in Canada to attend military school. In 1974, Lee Man Tat and Choi May Ling arranged for the three siblings to stay together in the place that Eddy Lee resided in Canada, to facilitate Choi May Ling visiting and taking care of them.

From 1975, Lee Man Tat's children again moved from the vicinity of Toronto to San Francisco. The couple had a lot of friends in San

The "adolescent five" who were still studying and growing, from left to right; Eddy Lee, David Lee, Elizabeth Lee, Sammy Lee and Charlie Lee.

Francisco, and they later bought a house in Belmont of the San Francisco Bay Area to facilitate their children's study in San Francisco. Both Charlie Lee and Sammy Lee spent their high school days in the Bay Area. According to Choi May Ling, when they moved from Canada to the States, she had to drive all the way from Toronto, down across the continent, to San Francisco in four days. At that time Eddy Lee was only 17 years old, while Elizabeth Mok was 16. Although the two did help to drive occasionally, it was mainly her that completed the journey. When Choi May Ling lived in California, she once drove out for shopping and got lost, however, she did not immediately rush to find her way back; instead, she continued shopping. Sammy Lee recalled that his mother was not worried at all and simply said: "Let's find the way out after we finished shopping!" He was truly impressed by the guts of his mother; just like his father, she remained calm and composed when facing danger or uncertainty.

Lee Man Tat himself was the eldest of his family, and he abided by the old adage of "eldest brother should act like father" and took care of all his siblings. He applied the same doctrine to his children and asked them to unite as a family to handle external affairs. Elizabeth Mok remembered that Lee Man Tat often taught her to "obey your eldest brother without question when you are away from home. You must follow his instructions, rightly or wrongly, without argument." It therefore followed that when Lee Man Tat and Choi May Ling were not with their children, their pocket money would be in the care of Eddy Lee for distribution to them. According to his siblings, Eddy Lee handled the money very well. Elizabeth Mok also mentioned that "Eddy would always keep the house clean and tidy, and wiped the floor every day. In our division of labour, I did the cooking, David washed dishes, while Eddy wiped the floor and disposed of the garbage... Eddy maintained a very high standard of cleanliness in the rooms." David Lee added that as a man of traditional mindset, Lee Man Tat expected his children to return to Hong Kong once or twice a year while they were studying abroad.

When the children were about to go to university, Lee Man Tat did not specify what courses they should take. But he kept on saying that "whatever you study, come back to help the family after your graduation." They all got his point. All of them selected majors relating to their family business. Eddy Lee graduated from the University of California (Davis) in 1980 majored in food science and technology, Elizabeth Mok also resided in food science. David Lee graduated from the Marshall School of Business at the University of Southern California, where he studied business administration and marketing.

The younger Charlie Lee and Sammy Lee studied in the renowned University of Southern California. Charlie Lee graduated in Chemical Engineering, while Sammy Lee graduated from the Department of Management and Finance. Sammy Lee was the only child who could have the leeway to work for Citibank in Hong Kong for over a year, while all his elder siblings were urgently "called back" by their father to join Lee Kum Kee upon graduation. The fourth generation shone in different positions in the family business later, and together, they have propelled Lee Kum Kee to a glorious level of development.

A Workaholic "Sprinter"

Entrepreneurs who are workaholic and energetic always tend to have a quick temper. They want to fulfill their visions and complete tasks as fast as possible, and place much greater emphasis on efficiency. Lee Man Tat obviously belongs to this type of entrepreneurs. When a decision is made, Lee Man Tat would rush to the goal with full power and speed. In this respect, Lee Kum Kee's dazzling speed of expansion from its humble plant in Wong Chuk Hang to the high-rise building of its office-cum-plant in Tin Wan from the 1970s to 1980s was a vivid example. Lee Man Tat was, at that exciting historical time, an unstoppable "sprinter."

Lee Man Tat's "sprinter" behaviours could first be seen in his style of driving. Chang Chin Vai, an old staff of Lee Kum Kee who has spent over 60 years with Lee Man Tat, mentioned that, from his early days in Macau, Lee Man Tat was famous for fast driving. Lee Shiu Nan once angrily commented that "when Man Tat accelerates his car, he took no notice of any passengers around." Lee Man Tat was fond of the Italian brand of Fiat exactly because its cars ran truly fast, especially when the red traffic light was turning to green, he could speed up quickly to leave other cars behind.

When shared with us his childhood days with his father, Eddy Lee mentioned an amusing anecdote concerning Lee Man Tat's driving style, which was never previously disclosed outside the family: "My father drove the family out one day. But when the car passed through Wong Nai Chung, he was provoked by the driver of a car on the adjacent lane. My father was so upset that he asked all of us to get off and raced with that guy by himself." The story revealed not only the "sprinter" character of Lee Man Tat, but his aggressive nature of never wanting to stay behind others.

Lee Man Tat is very proud of his leadership in advancing significant developments in the sauce and condiment industry since 1972, particularly concerning his achievement in promoting economical oyster sauce. He once boasted that "since 1972, others in the industry simply followed what I did." That said, Lee Man Tat was disappointed with his repeated failure in producing soy sauce since the 1970s. He even experienced a serious setback in the 1980s when he had to dispose 300 containers of soy sauce. When Lee Man Tat set up a Lee Kum Kee's production base in his ancestral hometown in Qibao, Xinhui in 1995, the company kept experimenting the production of soy sauce, and only came out with its own quality soy sauce in 2000. This demonstrated

The couple were overjoyed to see their children's educational accomplishments. The fourth generation of Lee Kum Kee graduated from prestigious American universities one after another in the 1980s, and the first amongst them were Eddy Lee and Elizabeth Lee, who graduated from the University of California, Davis.

the persistence of Lee Man Tat in pursuing his goal. It is indeed his conviction that soy sauce is the mostly consumed condiment and, without this as its product, Lee Kum Kee could never grow big and strong. This leads to another intriguing story that we will touch on in the following chapters.

Soon after Lee Man Tat assumed the stewardship of Lee Kum Kee, Lee Kum Kee needed to expand its production capacity due to the popularity of Panda Brand oyster sauce in the U.S. market. Lee Man Tat accordingly bought a piece of land in Wong Chuk Hang, Hong Kong Island and built a new two-storey plant for Lee Kum Kee. It was in fact a humble plant of only 7,000 square feet. During its time at Wong Chuk Hang plant, Lee Kum Kee, while busy in its production of economical oyster sauce, had begun to develop new products of condiment. Making soy sauce indeed was an important long-term target for Lee Man Tat, but he was occupied with his booming business at the time.

The Wong Chuk Hang plant was put into full operation in 1976. It is worth mentioning that Lee Man Tat bought that piece of land in Wong Chuk Hang from the Kadoorie family. Initially, the Kadoorie family was only going to rent it out. However, Lee Man Tat thought that it was a good piece of land which could facilitate Lee Kum Kee's development, so he approached the family directly for buying instead of renting the land and finally succeeded. His superb power of persuasion in this case impressed many.

Lee Kum Kee achieved a leap of its production techniques during its Wong Chuk Hang era. The company's previous workshop at No. 262 Queen's Road Central was still burning firewood for its oyster sauce production. As a result, whenever people boiled oyster extracts for the sauce, the workshop would inevitably become smoky, and the door must be kept closed. When Lee Kum Kee temporarily settled on Catchick Street, Kennedy Town in 1968, it started to switch from using firewood to diesel in stages. Lee Kum Kee began its steam production at the Wong Chuk Hang plant in 1976. Lee Man Tat accepted the suggestion of Choi Cheung Yiu (one of his brothers-in-law) to purchase steam boilers for the plant. They modified these steam boilers to suit the plant's setting and subsequently succeeded in adopting steam power to boil oyster extracts. Choi Cheung Yiu not only asked for additional equipment from Lee Man Tat, he wanted skilled talents as well. The easiest way, he proposed, was to head hunt people directly from Vitasoy's technical department, which had a plant just opposite Lee Kum Kee's. Lee Man Tat again took heed of the advice and consequently beefed up Lee Kum Kee's team of technicians to enhance its production capability.

Another major leap of Lee Kum Kee during its time in Wong Chuk Hang was the development of new products. This was mainly attributed to Choi Cheung Woon, another brother-in-law of Lee Man Tat. Choi Cheung Woon graduated from the Department of Veterinary Medicine of Lingnan University. He migrated to Hawaii in the 1950s and worked for Del Monte Foods in its quality control department. He was precisely the skilled talent that Lee Man Tat was looking for, thus he invited Choi Cheung Woon to join Lee Kum Kee. Chinese Marinade, one of the Lee Kum Kee's longstanding and popular products, was the creation of Choi Cheung Woon. When Elizabeth Mok graduated from the Department of Food Science in the United States, she once assisted her uncle Choi Cheung Woon to develop a series of new products at the Wong Chuk Hang plant. Elizabeth Mok never forgot her days in Wong Chuk Hang plant:

> *I particularly remember the process we came up with the first batch of new products, such as the Chinese Marinade and Char Siu Sauce. I needed to help in the powdering of Chinese spices, or boil them for their extracts... I still remember that when I went for a date after work, I smelt like Chinese Marinade, or more specifically I smelt like licorice*

or cinnamon, from head to toes, as if I was just chewing licorice in my mouth... Even after I had changed all my clothing of the day, my whole body was still exuding such smell.

Lee Man Tat's younger brother, Lee Man Lok, was also part of the collective memory of Lee Kum Kee's days in Wong Chuk Hang, He was then working hard in the plant as well. Both Eddy Lee and Elizabeth Mok mentioned in interviews that when Lee Kum Kee's Wong Chuk Hang plant began to operate, their uncle Lee Man Lok had also started to fully commit himself to Lee Kum Kee. He had a strong passion for the job, and was always the first to start work and the last to leave. Eddy Lee said that "he was harder working than anyone. If he didn't leave the plant, no staff would dare to go. He really loved his job." Eddy Lee however added that "Uncle Man Lok was an operator-type of person, often viewed things in details but could lose sight of the big picture." It could be said that Lee Man Tat tended to focus on the "big things" while Lee Man Lok preferred to handle "the details". The two brothers' perspectives were way apart. They could complement one another at work, but conflicts were inevitable.

Audacity in Bidding Land

It was only a few years after the full operation of the brand's Wong Chuk Hang plant that more space was needed, both its production and office. Lee Man Tat searched around for suitable lands as the company's new plant-cum-office, and eventually coveted a narrow plot in Tin Wan, Hong Kong Island. The land was only around 6,000 square feet, but it should be big enough to build a multi-storey building there for the company's production and office use. It was a piece of crown land released for the purpose of encouraging industrial development. An open land auction was accordingly set, and Lee Man Tat would need to compete with others for the land.

Inspecting product packaging at the Wong Chuk Hang plant in 1981.

On the day of the land auction, Choi Kai Yau, the "King of Frozen Shrimps" was the main rival for that piece of land. Choi Kai Yau already had a plant in the vicinity of the land, and had expressed his hope to acquire the land for synergy effect of land use. Choi Kai Yau was pretty much a counterpart of Lee Man Tat in many aspects. Both of them started their career early, and they were both aggressive in business, highly effective, extremely efficient, and reputed for their integrity in the business world. They were even of the same age. Choi Kai Yau earned his first fortune from the seafood business. He was Chairman of the Hong Kong Sun Hwa Group at the time, which dealt successfully with the Japanese in seafood (*Yi Zhoukan*, 2007). During the process of land auction, after several rounds of bidding, Lee Hung Cheong, bank representative from the camp of Lee Man Tat, became a little panic after sensing the strong will of Choi Kai Yau. He turned to Lee Man Tat and asked: "Mr. Lee, what do you think? They are real tough!" Lee Man Tat firmly replied with a concise message: "I am determined to get the land!"

In his recount of the fierce auction, Lee Man Tat recalled that in one round of bidding, he observed that Choi Kai Yau had twice privately consulted his aide, implying that he might become hesitant. Lee Man Tat therefore instructed Lee Hung Cheong: "Keep following each of their bid with ours quickly! Don't stop." Lee Man Tat was calm and focused, while Lee Man Lok, who was also attending the auction, was nervous and sweating. Apart from the two rivals the floor was silent while watching the fight. Choi Kai Yau finally gave up. He went over to Lee Man Tat to congratulate him, saying that: "Hi, we will soon be neighbours." It was a rather unexpected result to the public, as most thought that Choi Kai Yau would surely win the bidding. Gossips in the media on the following day highlighted "the victory of the King of Oyster Sauce against the King of Frozen Shrimps."

People gossiped about the event particularly because the auctioned price was much higher than market expectation. Lee Man Tat seemed to have acquired an expensive piece of land, but in the long run, the decision again had his foresight proven. The land appreciated merely two years after the auction. Moreover, Choi Cheong Too pointed out that conditions for the procurement of the land were in fact pretty good. The buyer did not have to make full payment immediately and could start to pay a few years later in installment, not to mention its lower rate of interests. That said, Lee Man Lok, a former accountant of Hang Seng Bank and a man meticulous in details, did not feel comfortable. He simply did not believe that acquiring a piece of land in exorbitant price with a loan was a good idea. Distance between the two brothers widened and paved the way for the family's second separation.

In 1982, Lee Kum Kee quickly completed its construction of a 17-storey building on that narrow piece of land in Tin Wan. Compared to the Wong Chuk Hang plant, the Tin Wan plant was not only bigger in space, but also equipped with a set of second-hand equipment acquired by Eddy Lee and his uncle Choi Cheung Yiu from the United States after a labourious search. The second-hand equipment came from a Chicago company which specialised in equipment for dairy production. Eddy Lee contended that it was cheap and good, and after some modification, greatly improved the production efficiency of Lee Kum Kee's oyster sauce.

Lee Kum Kee was expanding truly fast in the 1980s. Even though Lee Man Tat was highly efficient and effective, he might not foresee that in 1984, two years after the operation of Tin Wan plant, Lee Kum Kee needed more space for its development again. While this was welcoming to the entrepreneurial minded Lee Man Tat who was always aiming high in company performance, the under-current of crises was also sneaking in.

Conclusion

The vicissitude in the track of history is the litmus test for the wisdom of every momentous decision. Lee Shiu Nan was distressed with the first separation of the Lee family, but Lee Man Tat, who could always see the big picture from the small scene, realised the opportunity behind that troubled family division. He then seized the moment and turned crisis into opportunity. As a result, Lee Kum Kee not only remained intact but even flew higher with renewed momentum.

As we all know, pruning dead leaves and decaying branches from trees could prevent the spread of bacteria, as well as allow sunlight to penetrate through their canopy into the whole tree, ensuring better air circulation and facilitating the sturdy growth of the trees. From this point of view, the separation of the Lee family in the early 1970s might be a blessing in disguise for Lee Kum Kee. After eliminating internal contradictions and conflicts, the Lee family was able to conserve its energy for clearer targets. Only a few decades under the helm of Lee Man Tat, Lee Kum Kee excelled through a series of breakthroughs and rapid expansion. It was no longer the humble family workshop that it once was.

The Tin Wan plant and its huge advertising banner of Lee Kum Kee
Oyster Sauce in 1982. Both were rather imposing.

1   The notice was published, among others, in
    South China Morning Post on 17 July 1972.

2   According to Lee Man Tat, Jaxon Import
    & Export Company's counterpart in San
    Francisco was Wing Sins Chong Company,
    and the boss's surname was Siu. Jaxon's store
    in Hong Kong was then situated at Pottinger
    Street, Central.

3   Charlie Lee mentioned that Chinese
    restaurants in North America often used
    oyster sauce to cook beef. They sold "Fried
    Beef with Oyster Sauce" to the local white
    people and African Americans.

Ground-breaking ceremony of Lee Kam Kee's plant in the Tai Po Industrial Estate. A five-storey building would soon be completed on the plot, being both spacious and imposing. The plant marked a new stage of the development of Lee Kum Kee.

FLYING HIGHER AMONG CRISES

In 1986, my brother wanted to break up with me, and I finally paid him HK$ 80 million in cash. I was short of cash even after selling out everything. I sold out Brilliant Building, which was over ten storeys high, located on Queen's Road Central. The plant situated on No. 31 Wong Chuk Hang, next to Vitasoy's plant, was also sold on the cheap for a mere HK$ 11 million. As for the 17-storey building in Tin Wan, it was sold for only HK$ 46 million. The construction of Tai Po plant was scheduled to be completed in 1988, but I was simply running out of cash and needed to secure bank loans for it.

Lee Man Tat

INTRODUCTION

Hong Kong was again clouded with uncertainties in the 1980s when the Chinese and the British governments commenced their negotiations on Hong Kong's future. Hidden currents swirled in its socio-political scene, and the conflicting process of negotiations was, to say the least, elusive and perplexing. Even after the two governments confirmed that China would resume its exercise of sovereignty over Hong Kong on 1 July 1997 and signed the Sino-British Joint Declaration, differences and disputes remained in the ensuing negotiations on various arrangements over the transitional period. The local market and society suffered from such a precarious state of affairs, and the community at large was plagued by an enduring crisis of confidence. Investors tended to be over-sensitive and subsequently over-reacted to certain political events or gestures of politicians, which then led to a volatile market and an agitated society.

Amidst the ever-changing new developments, faint-hearted people who lost confidence in the future of Hong Kong sold out all their assets and packed up everything for migration, whereas Lee Man Tat, a veteran entrepreneur who embraced Hong Kong as his home and China as his motherland, was happy to see the end of colonial rule and, as always, foresaw a brighter future of the territory through his talent of seeing the big scene from the small one. More importantly, as Lee Man Tat could always seize the opportunities popped up in such panic situation of crisis and grasp what others discarded, he had set a bigger stage for and maximised the dynamism of Lee Kum Kee to scale new heights. Looking back, it was in fact a momentous turning point for Lee Kum Kee to escalate incredibly and left other rivals in the industry far behind.

Sino-British Negotiations and the Tai Po Plant

Lee Kum Kee's purpose-built factory at No. 40 Shek Pai Wan Road, Tin Wan started to operate in 1982. Although it was a high-rise building with 17 storeys, the plant was actually situated on a small and narrow piece of land, with rather limited space on each floor. Apart from fitting in the whole set of their modified second-hand mechanical equipment for production, the Tin Wan plant also needed to reserve considerable area for office use, and therefore significantly restricted its operational efficiency. In order to automatise the filling process of oyster sauce in the production, Lee Man Tat and Eddy Lee visited manufacturers in Germany for such equipment. Lee Man Tat was most impressed by the modern automatic equipment available in Germany and eager to buy these, yet was short of capital to do so. The floor area of the Tin Wan plant was simply too restrictive for effective operation. Eddy Lee pointed out that they were forced to piece together the machinery according to the limited space available, which meant a "bended production line" to compromise with the building layout, thus a less satisfactory operation of the production.

The plan of Lee Man Tat and Eddy Lee at the time was to utilise the top floors of the Tin Wan plant as offices and those floors below down to the seventh floor to be used to store packaging and production materials. Industrial production was confined to the six floors below. It started with the boiling of oyster extracts on the sixth floor, and the making of sugar syrup on the fifth floor. The fourth and the third floors were used for packaging, while the second floor functioned as warehouse. Lee Kum Kee's products would eventually be moved down to the ground floor for distribution; however, challenges arose at this final stage of operation. The small industrial area in Tin Wan did not have a well-planned gyratory road system, and the street in front of their factory's exit was pretty narrow. Even worse was the fact that a seafood processing factory of the Sunwah Group was located right opposite to its exit. The son of Choi Kai Yau, Choi Koon Shum,(1) who had just returned from his study in Japan, helmed the Sunwah Group and was equally aggressive in expanding themselves. Eddy Lee grinned as he recalled: it was a narrow street with a bend at the corner, and container trucks of both companies had to accommodate each other to park and load their respective cargoes, so much so that there was frequent delay in loading, even to the extent of affecting shipments. It was definitely a nuisance to both companies in their businesses.

That said, Lee Kum Kee's Tin Wan plant was larger than its Wong Chuk Hang plant. The new production base in Tin Wan finally allowed Lee Man Tat to produce certain new products which he was unable to handle in Wong Chuk Hang. Oyster sauce unquestionably remained its key product, but Lee Kum Kee Hoisin Sauce, Chinese Marinade, Black Bean Garlic Sauce, Chili Garlic Sauce, Char Siu Sauce etc. also made their debuts one by one in the market. These innovative new products were popular and promising, which soon entailed an even larger production base and office accommodation to catch up with their expanding sales and operation.

Under the helm of Lee Man Tat, Lee Kum Kee went through several periods of high-speed growth over the decades. It took off for the first time in the 1980s. Lee Kum Kee's worldwide revenue, for example, increased some 15-fold from the early 1980s to the mid-1990s. While the figure was merely several million US dollars in the early 1980s, it exponentially reached the mark of over US$ 100 million by 1994, and was doubling every three years by the late 1990s (Yim & Mak, 2005: 68). Lee Man Tat surely did not slow down his pace after Lee Kum Kee crossed over the mark of US$ 100 million in its turnover. He remained highly concerned on the numbers and requested his sons and the whole management team to keep up the company's vitality and momentum for expansion. Lee Kum Kee galloped to an even higher plateau at the turn of the 21st century when it again experienced a speedy

Lee Kum Kee adopted a proactive marketing strategy to attract restaurants and families in the 1970s and 1980s. Its posters highlighted Lee Kum Kee as a "good helper" to chefs and a "good friend" of housewives.

growth following the opening of its biggest production base in Xinhui, Guangdong. It was also a time when both its Sauce Group and Health Products Group established a firm footing in Mainland China.

Back in the 1980s, Lee Kum Kee's high-speed growth could be felt in many anecdotes around that period of time. For instance, in the early 1980s, in order to make inroads into unfamiliar new markets, Eddy Lee applied for a display booth in the world's largest and most significant food and beverage trade fair, Anuga (General Food and Drink Trade Fair).(2) He however failed to get a booth in the biennial fair because Lee Kum Kee was unknown to them. But it was during the decade of 1980s, according to Eddy Lee, that he took some 400 trips abroad for business as well as attending exhibitions. Such an amazing travel record not only reflected the super-rapid expansion of Lee Kum Kee's business, but also its aggressiveness in seizing new markets and the impressive coverage of its sales network worldwide. Lee Kum Kee's marketing slogan of "where there are people there is Lee Kum Kee" to a certain degree became reality by the late 1980s. And, indeed, Lee Kum Kee became a regular exhibitor of Anuga over the decade.

Lee Man Tat meticulously designed and planned for the space and functions of the Tin Wan plant. Obviously he thought the plant would be big enough for the operation of Lee Kum Kee for a good number of years. Lee Shiu Nan was also ecstatic over the completion of Tin Wan plant and, during its days of construction, was eager to visit the construction site on the arm of others, slowly and contentedly toured around this new base of his family business. However, against all expectations, the brand-new Tin Wan plant fell behind the needs of Lee Kum Kee's production in less than two years. By late 1983, Lee Man Tat had to ask Lee Man Lok and Eddy Lee to search for a new site to build another factory. However, options were limited, and the new Tai Po Industrial Estate, the first of its kind in Hong Kong completed in 1978, was their natural choice.

The father and sons regularly visited International expositions together to explore any improvement that Lee Kum Kee could apply to reach a world class status.

Tai Po Industrial Estate situated on a reclamation of 75 hectares. With Pat Sin Leng (Pat Sin Range) as its backdrop, the industrial estate is close to Tolo Harbour and just 1.5 kilometers away from the Tai Po New Town. Charlie Lee mentioned in the interview that Lee Man Tat had tried to apply for a site in Tai Po Industrial Estate in early 1982, and Eddy Lee was responsible for the feasibility study report. He seriously prepared a large number of documents for the application, but was disappointingly given the cold shoulder by the authority.

Eddy Lee vividly remembers that the then chief of Tai Po Industrial Estate, who was responsible for the approval of applications, said to him: "Mr. Lee, you are a manufacturer of oyster sauce, and do not fall within the high-tech industry. This is not the place for Lee Kum Kee." Eddy Lee felt deeply humiliated; however, unwilling as he might be to accept the "verdict", he could not but return to Tin Wan to continue holding the fort there. Nevertheless, later, there was light at the end of the tunnel with an unexpected twist in history, when Lee Man Tat immediately seized the moment and made an astute and incisive decision.

Before 1997, as a British colony, Hong Kong could be divided into three parts geographically, namely Hong Kong Island, Kowloon Peninsula and the New Territories. These territories fell into the hands of the British in different ways. Hong Kong Island was ceded by the Qing Government of China to the British in 1842, after she was defeated in the First Opium War and forced to sign the Treaty of Nanking with the British. The British again forced the Qing Government to cede Kowloon Peninsula to them in the Convention of Peking in 1860 following their victory in the Second Opium War. As for the New Territories, which accounted for 90 percent of Hong Kong's territory, it was leased by the Qing Government to the British as late as 1898 for a period of 99 years through the Convention for the Extension of Hong Kong Territory.

As the 99-year lease on New Territories would run out on 30 June 1997, the British Government was obliged to approach the Chinese Government for an extension in the late 1970s, starting with the visit of Hong Kong Governor Murray MacLehose to China in March 1979. On 24 September 1982, the same year that Lee Kum Kee applied for a site in Tai Po Industrial Estate, the British Prime Minister Margaret Thatcher visited China for the first time and met Deng Xiaoping, hoping to seek an amicable solution. The British initially insisted that all the three treaties related to their reign in Hong Kong were legitimate and valid, which meant negotiation for the future status of Hong Kong Island and Kowloon Peninsula to China was not necessary. The only issue, from their perspective, was the extension of the lease agreement concerning New Territories and North Kowloon. However, Deng Xiaoping was adamant on the issue of sovereignty, and insisted that it was non-negotiable. No consensus could therefore be reached from the first round of negotiation between the two governments.(3) A formal process of Sino-British negotiations commenced in July 1983, but the first four rounds of the talks came to a stalemate when the British Government proposed "an exchange of sovereignty for administration"(4) which was, however, firmly rejected by the Chinese Government. Again, no progress was achieved in the preliminary stage of the negotiations (Yuan, 1997; Ma, 2010).

Hong Kong was mired in a period of economic turmoil when the Sino-British negotiations showed signs of break down. The local community was particularly fretful as it had no right to participate in the negotiations and was beset by obscure messages and an uncertain future. In September 1983, which was around the fourth round of talks, Hong Kong dollar depreciated by 20 percent within a week. It fell dramatically from HK$ 7.9 to US$ 1 to HK$ 9.6 to US$ 1. Meanwhile, the Hong Kong stock market plunged, prices of goods soared, and people rushed to snatch up food items in the market. As a response to the currency crisis, the British colonial government urgently adopted a linked exchange rate system in Hong Kong on 15 October 1983, in which the HK dollar would be pegged against the US dollar at an internal fixed rate of HK$ 7.80 = US$ 1 to stabilise the HK dollar (Zheng & Luk, 2017).

Since then, the British conceded and agreed to withdraw from Hong Kong. Jardine Matheson (Jardines), a British conglomerate originally based in Hong Kong, announced on 28 March 1984 that it would relocate its head office (change of company domicile) from Hong Kong to Bermuda. The move was tantamount to a clear and most powerful vote of no confidence in Hong Kong's future, and sent shock waves across the whole commercial market (Zheng & Wong, 2006). After many rounds of negotiations, in September 1984, the two governments finally decided on the direction for Hong Kong's future, and thereby announced that Hong Kong would return to China on 1 July 1997.

The confirmation of Hong Kong's return to China prompted a significant number of pessimistic Hong Kong people to plan for migration. A disturbing trend of brain-drain emerged in the society and continued to escalate. Despite the unprecedented promises and commitments given by the Chinese Government to Hong Kong, including keeping the capitalist system "unchange for 50 years", "one country, two systems" and "Hong Kong people administering Hong Kong with a high degree of autonomy", Hong Kong business elites and the upper-middle class remained in general highly skeptical.

It was in the year of 1984 when people were in a state of panic that Lee Kum Kee's application to Tai Po Industrial Estate was revived. Situation changed drastically over the two years of Sino-British negotiations. Tai Po Industrial Estate was in serious difficulty at the time due to poor investment interests. Many big enterprises previously ready to move in abandoned their plans, faltering in their next steps or turning abroad. Paradoxically, it was the same chief of Tai Po Industrial Estate that called Eddy Lee back, and let Lee Kum Kee select any vacant plot within the industrial estate at will. Without hesitation, Lee Man Tat picked a plot best situated in the industrial estate which was at the junction of Ting Kok Road and Dai Fat Street, one of the main entrances into the industrial estate. According to Charlie Lee, this square piece of land was originally taken up by an internationally renowned pharmaceutical company for its new factory, but the company gave up in view of the highly uncertain future of Hong Kong, and hence the plot became available.

Eddy Lee still clearly remembers details of the purchase of land to this day:

> It was HK$ 74 for a square foot then. Because the plot was close to Ting Kok Road, we expected the cost of piling might be cheaper as the land was already there before reclamation. The area a little bit beyond might be reclaimed land, and you really didn't know what would happen while piling. It was a plot of around 120,000 square feet which could allow us to construct buildings with a total floor area of 330,000 square feet.

Timely grasping what others discarded in adverse circumstances and became winner later with improved situation might earn Lee Man Tat the acclaim of "astute decision-maker", but the implication was far more than that. Back to the historical context of Sino-British negotiations between 1982 and 1984, Lee Man Tat was in fact making a critical life-or-death decision for Lee Kum Kee. He threw in all the family resources to build another new factory in Tai Po while others were either sitting on the fence or simply moved abroad: if his decision turned out to be wrong, Lee Kum Kee might not have a second chance to survive. Lee Man Tat's decision was neither out of pure patriotism nor blind optimism, but based on his pragmatic judgment in seeing the big picture from the small detail. Most of his relatives and friends however did not see eye to eye with him. Amongst them, Lee Man Lok took particular exception to his decision. He was unhappy ever since Lee Man Tat bade for the land in Tin Wan with an exorbitant price. Discords between the two brothers in business judgment foreshadowed the second family separation of Lee Kum Kee.

Concurrent Capital and Health Crises

Referring to newspaper advertisements of the local supermarkets and products on sale in Hong Kong in the early 1980s, one could easily see that Lee Kum Kee had already successfully established an image of "the local (Hong Kong) quality products" as early as 1980s. It is nevertheless worth noting that the main contributors to Lee Kum Kee's rapid growth in the 1980s were from neither the Hong Kong market nor the Mainland China market, but its traditionally dominated overseas Chinese markets in North America as well as those in Europe and Southeast Asia which Lee Kum Kee was then vigorously exploiting. Throughout the 1970s and 1980s, Lee Man Tat still needed to frequently shuttle between Hong Kong and San Francisco, Los Angeles, Chicago, New York, Boston and other big cities in the United States. There was nevertheless a vast difference between his trips in the 1970s and the 1980s. Lee Man Tat mostly went alone from Hong Kong in the 1970s, while tapped into the local business networks of partners and friends in the United States; whereas in the 1980s, following his children's graduation from universities in California, his trips to the United States were always accompanied by them serving as his reliable and capable assistants.

Eddy Lee, Lee Man Tat's eldest son, graduated from the University of California at Davis in 1980. He was immediately called back to Hong Kong by a one-way ticket from his father. Lee Man Tat told him the company's business was booming but it was under-staffed; he had no alternatives but to ask Eddy back. Eddy Lee recalled that once returned, he was on his own to "deal with everything" across the board, "it's

The Tai Po plant in construction: a sign of opportunities for the advancement of Lee Kum Kee.

kind of a one-man operation." Apart from the headquarters in Hong Kong, Lee Man Tat also wanted to have his own man in the United States, and that heavy responsibility fell upon the shoulders of David Lee, who graduated from the Marshall School of Business, University of Southern California in 1982. Likewise, Charlie Lee immediately returned to Lee Kum Kee after graduating from the Department of Chemical Engineering, University of Southern California. He first followed his father to handle business in the United States and returned to Hong Kong in early 1986 to help promoting Lee Kum Kee's sales. The marching order from his father was also to "deal with everything." It was a time that China had already embarked on its economic reform and opening-up, thus Charlie Lee was also sent by his father to the Mainland to explore the new market.

David Lee got married in the United States right before his graduation and was the first amongst his brothers to build his family. David Lee recalled that he also joined Lee Kum Kee shortly after his graduation. While he had already found a job locally, his father suddenly called up and told him that Lee Kum Kee was facing a cash-flow crisis, and needed him to help. It was the bad debts of some of the company's distributors in the United States, which amounted to several million US dollars, and his task was to collect these monies back. David Lee was obliged to take up the task immediately. Apart from this, Lee Man Tat had long desired to have a more direct control of the America market; he therefore set up the first office of Lee Kum Kee (USA) Inc. in Los Angeles in 1983 with David Lee as the director, and Choi Cheong Too,(5) another pioneer of Lee Kum Kee's business in North America, to be the General Manager with special responsibility for sauce business in both North and South Americas.

Lee Man Tat also bought a warehouse in Monrovia located in the suburb of Los Angeles, as the first step to build Lee Kum Kee's own distribution network in the United States. David Lee had since been responsible for the United States market for nearly a decade until he returned to Hong Kong headquarters in 1991. During his days in the United States, David Lee also assisted Lee Man Tat to set up a new brand called "House of Lee" in an attempt to open up the market for non-Chinese American consumers. Products of the House of Lee remained Chinese-style condiments, yet with a high-end brand image as well as higher prices. Unfortunately, "House of Lee" was unable to consolidate its high-end brand image and failed. It was in fact priced out of the market and defeated by Lee Kum Kee's own products.

The young and inexperienced David Lee and Charlie Lee started their business careers in the United States with a not too glamorous job: collecting unsettled payment from Lee Kum Kee's distributors, large or small, in different cities. One of Lee Man Tat's best tactics in promoting business was the previously mentioned "selling with credit consignment", which meant he would send the goods to Lee Kum Kee's distributors for sale first, and collect payment after they had sold them out. But such a tactic was based entirely on mutual trust and integrity; if distributors defaulted in payment, or used it as a convenient way to support their cash flow, Lee Kum Kee's capital chain might risk rupture. Lee Man Tat may sound relaxed when he relates the case now, but in the 1980s, when he kept re-investing the company's capital to push for its aggressive growth, any mishandling might lead to catastrophic crisis. Thus, when the situation began to deteriorate, Lee Man Tat had to send his sons to personally collect payment from the distributors. It was in early 1985 that Lee Man Tat realized the business in the United States was in deep water. It could be fatal in the sense that Lee Kum Kee's capital chain rupture was imminent.

The trouble came from Ng Tung Hoi,[6] a major distributor of Lee Kum Kee in the United States. When Lee Man Tat gradually shook off influence of Kam Shan Chong on Lee Kum Kee, he attempted to build his own networks of sales in both the East Coast and the West Coast of the United States. However, as the country was not his home turf, it was almost impossible to ignore the established distributors in major cities. But the tricky part was, once he approached these local Chinese distributors, most of them requested to be the "general distributor" of Lee Kum Kee, which meant that they wished to have a monopoly on the distribution network of either the East Coast or the West Coast. According to David Lee, Lee Kum Kee's network formulation in the United States was pretty much in shape by the 1980s, with a major distributor on the East Coast and another on the West Coast respectively. And in terms of business turnover, the East Coast prevailed over the West Coast.

1985 was a very difficult year to Lee Man Tat. Earlier in mid-1984, his younger brother Lee Man Lok began to raise the distressful issue of family separation, which made him truly agitated and anxious. The serious problem caused by Ng Tung Hoi followed in early 1985 by which time this major distributor in the United States had still failed to honour his payment for over 180 days after importing a large amount of goods from Lee Kum Kee, much beyond the deal of settlement within 90 days. Lee Man Tat was extremely worried and went with Charlie Lee to New York to investigate. They were shocked by the scene: a total of 99 containers of their oyster sauce were still piling up in the dock's warehouse. Ng Tung Hoi neither distributed them to the market for sale nor made the payment. They rushed to Ng Tung Hoi's office for an explanation. According to Charlie Lee, Ng Tung Hoi was shameless enough to give them a poor excuse of "bad business environment" and even uttered that "if the Lee family is interested, kindly sell them by themselves."

After New York, Lee Man Tat and Charlie Lee continued their trip in the United States to monitor situations of other distributors. Perceivably, the embattled Lee Man Tat must be fret and deeply worried at heart. But as a gastronome, Lee Man Tat never stopped enjoying wining and dining throughout the trip, and was particularly fond of those bargain local delicacies like lobster, abalone and shark's fin. As a result, Lee Man Tat began to develop stomach problems when he was in Boston. Nevertheless, they continued on their way to Chicago and San Francisco; but by the time they reached San Francisco, Lee Man Tat's stomach-ache became unbearable. That night when they were about to take their next flight to the last stop in Los Angeles, Lee Man Tat took an X-ray in the hospital at the airport and subsequently took painkillers, barely able to take the flight with a view to inspect the newly purchased warehouse in the suburb of Los Angeles.

Choi Cheong Too, General Manager of the Los Angeles office, was at the scene when Lee Man Tat was struck by such an acute illness that almost claimed his life:

> *That morning, Lee Man Tat asked me to accompany his father(7) for a porridge breakfast. On our way to the restaurant, I asked Lee Man Tat: "Today is Saturday, do you have enough painkillers?" Lee Man Tat said there were still a couple left. I then suggested to bring him to the hospital after breakfast for a follow-up consultation, lest he was unable to find a doctor on Sunday. After breakfast, Lee Man Tat felt very uncomfortable. I took him to see the doctor, but there was a lot of people in the hospital. Given his emergency situation, I requested to jump the queue. When the doctor pressed his stomach, he was extremely painful. The doctor realised his critical condition and immediately arranged surgery for him. At that time, Lee Man Tat was already in state of semi-coma... When Choi May Ling arrived, the operation had been going on for five hours. It was indeed a long operation. When it was finally over, Choi May Ling and I treated the hungry Chinese doctor to a good meal.*

Charlie Lee was also at the scene. On the reason why his father was suffering from such acute stomach problem, Charlie Lee had this to say:

The next day, Mr. Choi(8) and I accompanied him to see the doctor.
Father was so painful that his face turned black. It was a case of bowel
perforation and he was immediately admitted for a surgery. After his
first operation, the hospital performed a series of specimen tests which
revealed that 11 kinds of worms were found in his stomach! He was such
a gourmand that even those "super Chinese worms" were there... He
went through a second operation in March (1985).

The acute colitis almost claimed Lee Man Tat's life should he miss an immediate surgery. He barely survived this serious health crisis in early 1985 with two surgeries in a row in Los Angeles and took the following four months for recuperation. Due to this unexpected illness, Lee Man Tat stayed in the United States for a stretch of nine months.

It seemed that greedy eating was not the only reason that had led to the bowel perforation of Lee Man Tat, so also was the impact of psychological pressure. During his days of recuperation, what was in the mind of Lee Man Tat? Did he, as in the 1960s when he was frustrated with his career in Lee Kum Kee and tried a sleep in coffin, reflect on his mortal life again? Lee Man Tat seldom shared with us what he did and thought in that specific period of time, but from various other sources of information, we could figure out that it was a different scenario this time. Lee Man Tat did take a rest in Los Angeles, but it was by no means a peaceful recuperation. He could not bear slowing down his pace while Lee Kum Kee was taking off, or being dragged down by internal strife and conflicts. Lee Man Tat decided to solve the problems one by one while in recuperation.

It was first of all, the tricky issue of Lee Man Lok pushing for family separation. With the help of his son-in-law Christopher Mok and the Mok family, Lee Man Tat approached the law firm Ko and Co. as his representative to prepare for the battle in court. Secondly, as far as the management of Lee Kum Kee headquarters was concerned, Lee Man

Lee Man Tat (1st right), Choi Cheung Yiu (2nd left), Choi Cheung
Woon (2nd right) and Eddy Lee (1st left) visited Yiu Hoa Produits —
Lee Kum Kee's distributor in Paris — in 1981.

Tat kept close contact with Eddy Lee in Hong Kong, issuing instructions and making decisions in Los Angeles, including the confirmation of oyster extracts' quality sent from Hong Kong as samples. As for the payment issue of Lee Kum Kee's major distributor in New York, he left it to Choi Cheong Too and Charlie Lee for a short-term solution on the one hand, while sped up the building up of Lee Kum Kee's own network as a long-term solution on the other. He subsequently set up the New York Office in 1986 in addition to the Los Angeles Office, and transferred the experienced Choi Cheong Too from Los Angeles to New York. It was the beginning of Lee Man Tat's new master plan in the United States.

There was one more incident worth mentioning during Lee Man Tat's recuperation in the United States. A low-key Jewish millionaire was interested in the brand of "House of Lee" created by Lee Man Tat and David Lee. He wanted to buy it over or seek a cooperation with the Lee family. After discussing with his sons, Lee Man Tat eventually turned down the offer, which might be a golden opportunity for "House of Lee" to be developed to the same height as Lee Kum Kee in the international market. Lee Man Tat's consideration, as Choi Cheong Too observed, was his strong preference "to command the direction and operation of his own company," and the fact that he was never willing to take orders from others. It is thus understandable that persistent in his mindset, years later, when Lee Kum Kee has become much bigger, Lee Man Tat still never contemplates the listing of Lee Kum Kee in the stock market.

In October 1985, Choi Cheong Too and Charlie Lee were asked to go to New York again for discussion — or more specifically, confrontation — with Ng Tung Hoi. Charlie Lee recalled that Ng Tung Hoi warned them not to interfere those stores in his New York "territory", they therefore visited other cities in the East Coast, and "tried to figure out the real situation of Lee Kum Kee in these cities one after another. Was our oyster sauce really unpopular in these cities? And where were our potential customers?" After they had ascertained the actual sales of Lee Kum Kee's oyster sauce in the East Coast, Choi Cheong Too and Charlie Lee were confident enough to take back the stocked goods from Ng Tung Hoi in stages, then stored them up and distributed them to the market later. However, since the products had been left in the dock's warehouse for too long, bringing them out entailed a large amount of storage fee, which was even higher than the value of the goods. Lee Kum Kee indeed suffered substantial loss from the case.

When Lee Kum Kee commenced its operation in New York, it had no office. It was registered under the address of an apartment unit. Charlie Lee was still amused when he recalled the awkward moment of interviewing an applicant for the office secretary with Choi Cheong Too in the apartment unit: "A lady came for the job, but when she opened the door, she became hesitated. Eh? It's a residential unit? Isn't it an office? Please wait, let me go back to ask my husband... What kind of secretary do you want to hire? Why do you need a secretary in an apartment? (laughter) You see, we really started from scratch."

After parting with Ng Tung Hoi and bringing back a large stock of goods, Charlie Lee and Choi Cheong Too needed a warehouse of their own to store the goods. They found one in the city's Brooklyn borough, and bought it from the Jews. While working in the office of the warehouse, Charlie Lee recalled that when the truck drove in, he often saw the staff "strangely busy doing something at the truck's front. After that, something might just disappear. It could be a missing chair today, and something else tomorrow. It turned out that they were stealing! In some cases, two guys separately worked on the truck's front and rear parts. New York was no doubt a financial centre in daytime, but a totally different environment at night. The world changed dramatically from around 4 or 5 pm..." In any case, they accomplished their mission, and assisted Lee Man Tat to finally complete his strategic sales operational plan for both the East Coast and the West Coast of the United States.

Second Family Separation amidst Internal and External Difficulties

Lee Man Tat spent months in Los Angeles for his recuperation from the surgeries. The moment of his recovery met the birth of his eldest grandson, Brian Lee, in Toronto, Canada. It was not only a joyous but a highly significant moment to Lee Man Tat. Apart from the fact that he had become a grandfather, it also marked the beginning of the fifth generation to Lee Kum Kee, which was indeed a great event from the perspective of Chinese culture. Therefore, despite the fatigue of travelling, Lee Man Tat flew to Toronto to visit his newborn grandson before anxiously returning to Hong Kong. Awaiting him in Hong Kong was not just the uncertain local environment of business, but a lawsuit with his brother which he loathed to face.

As mentioned in earlier chapters, Lee Shiu Nan and Chan Choi Kam raised a total of two sons and six daughters. Lee Man Lok ranked fourth among the siblings and was much younger than his elder brother Lee Man Tat, with a gap of more than ten years between them. According to David Lee, Lee Shiu Nan was a traditional man who preferred sons to daughters, thus his daughters had no role at all in Lee Kum Kee. In contrast, he cherished the two sons and entrusted them with heavy responsibilities. In 1972, following the first family separation of Lee Kum Kee, Lee Man Tat helmed the company, while Lee Man Lok resigned from Hang Seng Bank and joined Lee Kum Kee as well shortly after. At that time, Lee Shiu Nan had fully retired, thus Lee Kum Kee was essentially managed by the two brothers.

Lee Man Tat apparently loved and cared about his younger brother very much. In the early 1950s, when Lee Man Tat was running his Tat Sang Leatherware Company in Macau, Lee Man Lok lived in the shop of Tat Sang. After graduating from Lingnan Middle School in Macau, Lee Man

Lee Man Tat and Eddy Lee visited the Wing Yip Company — Lee Kum Kee's distributor in Manchester, UK — in 1981.

Lok attended St. Stephen's College in Stanley, Hong Kong, and it was again Lee Man Tat who took care of him. Upon his graduation, Lee Man Lok's first job at Hang Seng Bank was introduced by Choi May Ling. He later married a younger sister of Sinn Man Yee, the boss of Pastelaria Hin Kee in Macau, and Choi May Ling was their matchmaker, as Sinn Man Yee's wife was her former classmate. In 1982, when Lee Man Lok was diagnosed nasopharynx cancer, it was again Lee Man Tat and Choi May Ling who enthusiastically sought the best doctor for him and arranged his treatment in the Hong Kong Sanatorium & Hospital for his recovery. Lee Man Lok was passionate with his work in Lee Kum Kee, as confirmed by all the children of Lee Man Tat. Their Uncle Man Lok was hardworking and attentive to every aspect of the business, much more diligent than any staff of Lee Kum Kee. Lee Man Lok, from this perspective, seemed to have no reason to ask for a break up.

And yet, Lee Man Lok raised the sore issue of family separation in 1984, which was totally out of Lee Man Tat's expectation, and shocked both the family and Lee Kum Kee. On a closer look at the causes of the brothers' break-up, one could see that it was basically the fundamental differences of their values and personalities, which were elements truly difficult to reconcile. Until today, Lee Man Lok's press for a separation—particularly his legal action to seek a winding-up of Lee Kum Kee — still grieved Lee Man Tat and Choi May Ling. However, if we look from Lee Man Lok's perspective, his motivations for a break-up are not hard to be comprehended.

Lee Man Lok's diligence was obvious to all. But compared to Lee Man Tat's obsession with business, the two brothers were markedly different in style and manner. Elizabeth Mok mentioned that Lee Man Lok, since the days in Wong Chuk Hang plant, was often the earliest to go to work and the last to return home. Eddy Lee nevertheless noticed that his industrious Uncle Man Lok had a tendency towards too attentive to details and overlooked the overview. For example, the brand labels on Lee Kum Kee's products were manually stuck at the time instead of automatically processed by machine. Lee Kum Kee only had a small number of staff at Wong Chuk Hang plant, who occupied two round tables during lunch time, and the table that was used for labelling was just next to the dining tables. "When it was time for lunch, the staff began to gather around dining tables, but Uncle Man Lok didn't eat and sat on the table next to them, and continued working on labelling. The staff were uncomfortable in such awkward situation and were forced to come to help, thus delaying the meal."

Lee Man Tat was starkly different. He preferred to look at big pictures, and loved to talk about big things, big issues and grand strategies. If he thought a business's overall direction was right, and that he had 60 to 70 percent of confidence that it could be done, he would take the risks and go ahead, while leaving details of implementation to a later stage. David Lee hit the nail right on the head when he pointed out that conflicts between his father and his uncle were mainly due to their differences in investment and business styles, which could be summarised as "aggressive" against "conservative". Lee Man Tat was very proactively forging ahead and always re-investing in the full adventurous spirit of an entrepreneur. Lee Man Lok, on the other hand, was very conservative in terms of investment. For instance, he disliked the idea of borrowing a huge sum of bank loans for investment or the building of new factory as, first of all, it was too risky to him, and secondly, it was not worthwhile in view of the long-term payment of loan interests.

With his conservative financial outlook, Lee Man Lok naturally could not accept Lee Man Tat's audacity in bidding that piece of land in Tin Wan with an exorbitant price, not to mention that only two years after the operation of Tin Wan plant, Lee Man Tat dared to sign with Tai Po Industrial Estate for a huge investment of HK$ 30 million to build another factory at a time of high uncertainty in Hong Kong's future. The two brothers' business styles were also different. Lee Man Tat always loved to socialise and entertain with generous spending, and that was in fact a key element of his success in business. It is nevertheless worth noting that Lee Man Tat's generous spending had also become an issue of discord with his cousins during the first family separation. Choi Cheong Too mentioned that Lee Man Lok was prudent on the company's financial spending, and apparently had issue with Lee Man Tat's lavish spending in social networking.

Certain internal and external troubles also prompted a psychological change of Lee Man Lok that finally led to his insistence on a break-up. As described by Eddy Lee, the internal cause was the diagnosis of his nasopharynx cancer in 1982. Lee Man Lok once thought that he could not survive. Even though he was cured later, he remained worried about a relapse. Moreover, all his four children were still young at the time,(9) unlike the five children of Lee Man Tat who were all grown-ups, graduated from universities in the United States one by one and joined Lee Kum Kee. In this regard, Tsoi Cheong Ming, Choi Cheong Too, Eddy Lee and Elizabeth Mok all mentioned that Lee Man Lok's wife was especially anxious and had a deep sense of insecurity, fearing that if Lee Man Lok finally succumbed to the cancer, she could not make it on her own.

Lee Man Lok might also intrinsically feel a sense of loss when Lee Man Tat's children began to join Lee Kum Kee. Before that, it was always his job to help his brother to read English documents of the company, yet when Eddy Lee came back to Hong Kong in 1980, it was Eddy Lee that took the task from him. "He felt he was no longer needed." Worse was the simultaneous impacts of external troubles: the day Lee Man Lok was diagnosed with nasopharynx cancer was also the time the British and the Chinese governments kicked off their formal negotiations. Lee Man Lok and his wife had no confidence at all in Hong Kong's future after the handover and, like so many other contemporary Hong Kong people, intended to migrate to Canada. It was therefore a rational choice to Lee Man Lok in seeking a family separation as soon as possible before Lee Man Tat threw in all the company's resources into Tai Po Industrial Estate to build his new factory. The couple could then acquire a hefty sum of money that they believed they deserved, and after that, they could get away from Hong Kong to start a new life in Canada.

However, Lee Man Tat loathed the idea of family separation in his heart, as he always believed that "family harmony" was the root of prosperity for family business, not to mention that he still had a vivid and unpleasant memory of the first family separation in the early 1970s. Moreover, it was the worst timing as Lee Kum Kee was rapidly expanding and urgently needed capital for re-investment. Lee Man Lok indeed could have a good share of the company's assets, but the separation would probably ruin this ancestral legacy. The most heartbreaking to Lee Man Tat was Lee Man Lok's legal action to seek the winding-up of Lee Kum Kee on the advice of his lawyers. It was, to Lee Man Tat, a painful betrayal by his beloved brother.

Lee Kum Kee's second family separation spanned from 1984 to 1986, bustling for nearly two years. During the court case, both sides spent a great deal of money to hire reputable lawyers for the lawsuit. It fell into a vicious cycle and put both families at loggerhead to the core. Eddy Lee said his father "suffered a great deal" from this event, and subsequently fell ill in the United States in 1985. "It was a very exhausting affair. We had little energy left for our business." At that time, the aged Lee Shiu Nan was still alive and saddened by the family conflicts. He did attempt to persuade Lee Man Lok not to leave the family.

It was said that tears welled up in Lee Shiu Nan's eyes when he gave advice to his son. Lee Man Lok was at a loss to reply at the time, apart from insisting on family separation. In Chang Chin Vai's memory, for the decades he had served the family, it was the third time Lee Shiu Nan shed his tears. The first sad moment for Lee Shiu Nan was related to the first family separation when Lee Shiu Nan failed to persuade his sister-in-law (Lee Shiu Dang's widow) to refrain from family separation. The second time was in 1976, when Lee Shiu Nan placed flowers on the body of his beloved wife Chan Choi Kam lying in the funeral parlour.

Today, details of the family separation that happened decades ago are no longer important. Simply put, Lee Man Tat and his sons were represented by the law firm Ko and Co., while Lee Man Lok was represented by Liu, Chan and Lam. Legal teams from both sides were equally powerful. Lee Man Tat spent lots of money to seek help from three Queen's Counsels led by the heavyweight Sir John J. Swaine,(10) while Lee Man Lok also invited two Queen's Counsels to his team.

Lee Man Lok stated in the court documents that since 1984, he had raised his objection against the deal between Lee Kum Kee and the Hong Kong Industrial Estates Corporation for the purchase and development of a site at Tai Po Industrial Estate, which was, in his view, beyond the means of the company and would involve a capital commitment of about HK$ 30 million. He thus requested the winding-up of Lee Kum Kee lest his interests be adversely affected.(11) The two sides continued to confront and argue, including debates on value of company assets and details of auditing. Lee Man Lok visited Lee Kum Kee's office in Tin Wan at times with accountants, and thoroughly checked the accounts books. But every time when he came, Lee Man Tat and his sons would shun to other places, and leave it to Ho O Shang, a staff from the finance department to deal with them. And according to Ho O Shang, once the check was over, he would go to Lee Theatre in the Causeway Bay to brief Lee Man Tat, so that Lee Man Tat knew what his brother was up to.

This civil lawsuit of family separation eventually reached a settlement in April 1986 through the persuasion of the judge. Lee Man Lok reluctantly accepted the division of shares on the basis of 40% entitlement for him and 60% for Lee Man Tat,(12) instead of his original demand of a 50 – 50 split. Lee Man Tat was willing to buy that 40% shares from Lee Man Lok at a premium of HK$ 80 million. Vincent Ko pointed out that Lee Man Tat, at that time, was facing an enormous financial pressure due to the construction of Tai Po plant. He could at best pay Lee Man Lok in instalments, just like what he did in 1972 during the first family separation. The two families parted their ways thereafter. Lee Man Tat and his sons arduously maintained Lee Kum Kee, while Lee Man Lok migrated to Toronto with his family. The couple moved back to Hong Kong to settle years after.

In order to raise a huge sum of HK$ 80 million for Lee Man Lok, Lee Man Tat was forced to sell out his quality properties in Hong Kong on the cheap in 1986, which was a buyers' market in real estates at a time when waves of emigration emerged. He could still feel the pinch today. During his interview with us, Lee Man Tat grumbled that:

> In 1986, my brother wanted to break up with me, and I finally paid him HK$ 80 million in cash. I was short of cash even after selling out everything. I sold out Brilliant Building, which was over ten storeys high, located on Queen's Road Central.(13) The plant situated on No. 31 Wong Chuk Hang, next to Vitasoy's plant, was also sold on the cheap for a mere HK$ 11 million. As for the 17-storey building in Tin Wan, it was sold for only HK$ 46 million. The construction of Tai Po plant was scheduled to be completed in 1988, but I was simply running out of cash and needed to secure bank loans for it.

Years later, when Elizabeth Mok happened to bump into Uncle Man Lok in Central, she would still go forward to greet him and chat for a while. A sister of Lee Man Tat came back from Toronto in one year and wished to orchestrate a reconciliation, hence arranged a meal gathering at Lee Man Tat's home and invited Lee Man Lok to join. Elizabeth Mok remembered Lee Man Lok did turn up:

> *Everyone was fine ostensibly. I remember uncle told my daddy that might be it was better for Lee Kum Kee without Lee Man Lok. If he were still in the company, he might hinder the development of the company. I think what he wanted to say was: they are brothers, if they have different opinions, he might stand in the way of daddy. Daddy wanted to build plants in Mainland China at a time when many people were migrating, thus Uncle Man Lok would not be amenable to throwing money into the Mainland project.*

Tortured by the impacts of two family separations, it is understandable that Lee Man Tat still cannot get over the threat by his brother to wind up the family business. This is a reflection of his dedication to business and sense of mission. He nevertheless stayed calm and did not allow emotions of indignation and resentment to rule over his head. Seeing through opportunities in drastic socio-political changes, Lee Man Tat finally mastered the situation and embarked on his journey to a higher plateau.

Flying Higher in the Centenary

From 1984 to 1986, Lee Man Tat was concurrently hit by a fatal health crisis and a capital crisis in the United States, worse was his brother's press for family separation around this critical time. He steadily solved the problems one by one, bringing Lee Kum Kee back from the verge of winding-up. When Lee Kum Kee's Tai Po plant was finally completed in 1988, Lee Man Tat was not only burdened with tens of millions of bank loans, his company was in "negative equity," which implied a highly precarious situation for Lee Man Tat and his sons: if only they had problems with their cash flow, they might go bankrupt. However, as a sturdy captain with a dominant leadership style, Lee Man Tat again benefited from the simplified "unified leadership structure" of Lee Kum Kee. The family separation ended with the departure of Lee Man Lok and his family, thus it was effectively a return to the similar situation in 1972 after the first family separation. Where to go and what to do next? Lee Man Tat could give the marching orders and take the family and Lee Kum Kee moving forward with him. Nobody could divert him from his course now.

The Tai Po plant had its groundbreaking ceremony in the winter of 1985, and began its construction in 1986. Lee Shiu Nan, an octogenarian, was unable to walk at the time. Family members would drive him to Tai Po for an eager look at the progress of the construction. Regrettably, he passed away in 1987, shortly before the completion of the plant. To remember his father, Lee Man Tat set aside a space in the entrance lobby of the Tai Po plant and placed a white stone statue of Lee Shiu Nan at the centre of the site. In 1995, he personally penned the inscription of the *Epitaph of Mr. Lee Shiu Nan*, and set the plaque beside the statue of Lee Shiu Nan.

1988 was the centenary of Lee Kum Kee, Lee Man Tat therefore purposely selected the auspicious date of 8 August 1988 to produce the first bottle of oyster sauce from its Tai Po plant. At the time of its 100th anniversary in 1988, Lee Kum Kee had reached a staff number of 108 people, which could already be seen as a modest achievement compared to its workshop days in the 1960s when it was only staffed by less than 20 people. But to Lee Man Tat, Tai Po plant was just another starting point for Lee Kum Kee. He was ambitious enough to make it a mere base for the next big stride to China and the world. Lee Man Tat's children, at this point, reminded their father that Lee Kum Kee should timely carry out institutional reforms as part of its modernization, as well as reshape its image.

Lee Kum Kee's trademark at the time was originated from Lee Man

Tat's amateur design in the 1960s. Lee Man Tat himself also realised that it was outdated. Since it was the centenary of Lee Kum Kee, Lee Man Tat decided to re-design its trademark. He accepted the advice of David Lee, who was professionally trained in marketing and based in the United States then, to seek the professional design service of S&O, a reputable company in San Francisco which designed corporate images for big American companies such as United Airlines and Citibank, thus charged a high price. To the surprise of Eddy Lee and his brothers, their father agreed to spend more than US$ 200,000 for a simple logo design of Lee Kum Kee trademark, which was in fact an investment for its intangible asset. The eye-catching trademark is everywhere today, as shown on every product of Lee Kum Kee. It looks like a rising sun shining upon a bridge, but the true implication of the trademark is to show how Lee Kum Kee accommodates Chinese culinary culture with the West's, as well as its goal in pushing forward its business from small to big.

After settling the court case of family separation, the Lee family was able to focus on its core business of sauce and condiments again. According to Ho O Shang, who worked for Lee Kum Kee's finance department at the time, Lee Man Tat sternly demanded his sons and the management team to achieve a doubling of Lee Kum Kee's business every three years. It was certainly not an easy target to meet by any means. Ho O Shang added that shortly after they moved into the Tai Po plant, Lee Man Tat frequently held "crisis meetings" to put pressure on them to close the gap between the upgraded production capacity and sales volume. His sons were fully involved in the process.

After a speedy expansion of Lee Kum Kee in the markets of North America and Europe throughout the 1970s and 1980s, Lee Man Tat was clearly aware that if he wished to maintain a super-fast growth of doubling turnover every three years, he must enter the world's most promising mega-market of Mainland China, which had just

Lee Man Tat and his four sons during the inauguration ceremony of
the Foundation Stone placed near the entrance of the Tai Po plant in
1988.

experienced the early days of her economic reform and opening up. As for the question of how to get started in this new market, he needed to think through thoroughly while taking cautious steps to explore the opportunities.

Conclusion

In the course of advancement of any society, enterprise and family, obstacles are unavoidable. If leaders are intimidated in the face of difficulties and challenges, or make poor judgments in the process, the society, enterprise or family concerned would pay a heavy price: they would either get stalled in their development, or fall into great danger. A leader's ability, vision and determination are therefore crucial to his institution's progress and fate.

The 1980s was a critical decade to both Hong Kong and Lee Kum Kee, which could be regarded as a landmark decade in their histories. It was a period frequently intertwined with crises and opportunities, thus uneasy for anyone to shun dangers and grasp opportunities and move forward. Lee Man Tat responded ingeniously in terms of the precision and timeliness of his decisions throughout the period, thus allowing Lee Kum Kee to scale new heights. In retrospect, these critical actions and decisions might appear easy or simple to take, but if placed in that historical context of the 1980s, the circumstances were very confusing, chaotic, and beset by various internal and external difficulties. In short, it was a decade of great uncertainty when those involved could feel helpless and unsettled. Yet, the courageous Lee Man Tat firmly steered Lee Kum Kee through that precarious ocean of the 1980s, which might otherwise capsise and disappear beyond the face of the earth. Without such leadership, he would not be able to reminisce over those trying years with the sense of joy and satisfaction that he now feels.

The plant at Tai Po Industrial Estate was completed in 1988.

1   Choi Koon Shum was only 25 years old when he took the helm of Sunwah Group from his semi-retired father Choi Kai Yau in 1982.

2   Anuga is an acronym of "Allgemeine Nahrungs-und Genussmittel-ausstellung" in German language, which is held biennially in Cologne, Germany.

3   Deng Xiaoping, referring to the issue of sovereignty, stressed that "China has no room for manoeuvre on this issue. Frankly, the issue of sovereignty is not an issue to be discussed" (Deng, 1993: 12). Nevertheless, the Chinese Government expressed a high degree of flexibility on the governance of Hong Kong once resumed her sovereignty over Hong Kong. Hong Kong would be allowed to maintain its independent economic and political systems for 50 years and managed by people of Hong Kong, with a high degree of autonomy. The same set of policies would also be applied to Macau and even Taiwan in the future.

4   It referred to the British Government's recognition of the sovereignty of China over Hong Kong in exchange for the British rights to continue to govern Hong Kong.

5   Choi Cheong Too is Choi May Ling's younger brother. He graduated from Far East Training School and worked for the Hong Kong Aircraft Engineering Company in the 1960s. Choi Cheong Too migrated to Hawaii in November 1968. He ran a small store there and used to act as a distributor of Lee Kum Kee's products. Lee Man Tat invited him to join Lee Kum Kee in 1982, and let him take charge of the Lee Kum Kee Office in Los Angeles in the following year. In 1986, he moved to the East Coast and headed the New York Office as General Manager, responsible for developing the market in the Eastern United States. Meanwhile, his elder brother Tsoi Cheong Ming was appointed to take up his post of General Manager in the Los Angeles Office.

6   This is a pseudonym.

7   Lee Shiu Nan was also in Los Angeles at the time.

8   Charlie Lee calls Choi Cheong Too "Mr. Choi" instead of uncle, and this is in fact a common practice in Lee Kum Kee. The brothers also address their father Lee Man Tat as "Mr. Lee." As a multinational family enterprise, Lee Kum Kee has many employees, including members of the Lee family and their relatives, thus the enterprise has adopted a modern personnel management system. Under the system, rewards and promotions are based purely on one's performance. The non-personal

addresses amongst Lee Kum Kee's staff reflect a high degree of institutionalization of the enterprise.

9   Lee Man Lok and his wife have two sons and two daughters.

10   Sir John J. Swaine was of Chinese-English parentage born in Shanghai. He graduated from The University of Hong Kong and the University of Cambridge, became a barrister in Hong Kong in 1960, and was appointed a Queen's Counsel in 1975. Sir John was reputed for handling civil cases. He was also active in politics, and was appointed an unofficial member of the Legislative Council in 1980. In 1993, with an election amongst the Legislative Councillors, he was made the President of the Legislative Council. Sir John left the Legislative Council in 1995, and was subsequently knighted by the Queen. Before retiring from legal practice, he was the most senior practitioner on the Hong Kong bar list. He died in 2012.

11   Affidavit Verifying Petition, 29 October 1985, Companies Winding Up (No. CWU302 of 1985), Supreme Court of Hong Kong.

12   Affirmation of Eddy Lee Wai Man, 7 April 1986, Companies Winding Up (No. CWU302 of 1985), Supreme Court of Hong Kong.

13   Brilliant Building was located at No. 194 Queen's Road Central.

Lee Man Tat and Zhao Yunhong, former President of the First Military Medical University, signed an agreement in 1992 to co-found the "Nanfang Lee Kum Kee Company Ltd".

INVEST IN CHINA AMIDST CRISES

President Zhao put forward the names of their research outputs, which I had totally no idea. I thought these were promising, thus we shook hands in less than 60 minutes and confirmed our intention to collaborate. That's the way the business began, and today, its development is far beyond my expectation.

Lee Man Tat

INTRODUCTION

The tortuous course of advancement of modern China had caused great suffering to Chinese people, and led to the dislocation of many of them over generations. One of their means for survival was to leave their hometowns for living abroad. According to informal statistics, in the Commonwealth countries alone, the number of overseas Chinese has reached 15 million (Wang, 2003).(1) Despite numerous humiliations, hardship and even fatal threats, overseas Chinese miraculously achieved a great deal in the business world through their toil, tears and sweat as well as their spirit to help one another in foreign lands. They have become an important force of economic construction in their hosting countries.

A common practice of the overseas Chinese that marvelled the world was the fact that they kept remitting savings back to their hometown to support parents and raise their children, as well as rebuilding their motherland (Hicks, 1993). Their homeland nostalgia also created a huge overseas market for Chinese products of different specialties. The rise of Lee Kum Kee's oyster sauce was precisely against such historical backdrop, which finally made it a popular condiment amidst the overseas Chinese, appeasing their nostalgia with a distinctive taste from their homeland. As for the "homecoming" mentality deep-rooted in their cultural genes, which entailed returning their due to their homeland once achieving something abroad, it prompted Lee Man Tat to likewise return to his ancestral hometown and motherland for investment. It was nevertheless a precarious journey of return, since during the early stage of China's reform and opening-up, most of the institutions and regulations were not yet fully established, and the investment environment was elusive. Even a little carelessness might lead to a sudden fall into the abyss, and fatally impacted on the family business.

Exploring Opportunities in the Reforming China

In the early days of New China, things mostly remained unsettled and were subjected to unexpected drastic changes. It was only after the pronouncement of "Reform and Opening-up" policy by the Communist Party's leadership headed by Deng Xiaoping in 1978 that China began to set in a rather different course which has since brought long-term economic well-being and political stability to the Chinese people. The momentous implications of such a change might not be felt by ordinary people then, yet the astute and far-sighted Lee Man Tat sensed the big change and, as always, by seeing the big picture from the small detail, snapped the opportunities. On the one hand, he bore hopes for the future development of China; while on the other hand, he started to think about some early steps in bringing Lee Kum Kee back to the Mainland market.

From the investment point of view, he realised that Hong Kong enterprises could take advantage of the very cheap land, labour and other elements of production at the time by moving their production lines and factories to the Mainland, thus reduced the costs of production and enhanced their competitiveness. And from the market point of view, he clearly foresaw the potential market forces that more than 1 billion people could bring: so long as they each bought a bottle of oyster sauce or any other Lee Kum Kee's product in a year, it was already a tempting number.

Lee Man Tat surely did not focus on business interests alone, his patriotic sentiment was deep and strong. He was born and raised in Macau, even though he did not suffer directly from the war fires of World War II, he belonged to the generation who grew up in the turbulent years of 1930s and the 1940s, profoundly impacted and influenced by the Chinese nationalist passion culminated in the hardship of the Anti-Japanese War. However, the end of World War

II swiftly followed by the Chinese Civil War, and when the dust had finally settled in 1949, Lee Man Tat left Guangzhou — and literally also left Mainland China — for nearly three decades. Until the end of the Cultural Revolution in China and Deng Xiaoping's comeback for economic reforms, Lee Man Tat essentially did not have any direct contact or interaction with Mainland China, including interaction with his two ancestral hometowns of Nanshui, Zhuhai and Qibao, Xinhui. His first visit to Nanshui was supposedly as late as 1983, while his first trip to Qibao was in 1990 after passionately invited by the local cadres.

The commencement of reform and opening-up in 1978 was, to Lee Man Tat, a watershed moment with far-reaching implications. Although he spent most of his time on business trips to the United States and Europe throughout the 1980s, he apparently kept a close eye on the new development of Mainland China, and cautiously conducted small-scale attempts to test the water of Mainland Chinese market. From the mid-1980s, when he could better grasp the situation, he began to push through a blueprint of aggressively building factories for production and sales in the Mainland with his children. Nevertheless, the process stalled in 1989 and the years that followed due to uncertainties arising from chaotic political turmoil besetting China at the time. There was thus a few years' delay until the spring of 1992 when Deng Xiaoping, through his tour to southern China, reinvigorated the process of reform and opening-up.

On the other side, since the early 1980s when China's domestic politics had already settled, local units of the Chinese People's Political Consultative Conference (CPPCC) and institutions in charge of overseas Chinese affairs attempted to approach Lee Man Tat. Documentary evidence suggested that shortly after Zhuhai CPPCC conducted a survey to ascertain actual situations of the overseas Chinese communities originated from Zhuhai for establishing contact, they discovered the connection between Lee Man Tat and Nanshui,

Zhuhai. In 1988, Zhuhai CPPCC wanted to nominate Lee Man Tat as its member, but the offer was politely turned down. The *Wenshi Ziliao (Humanity and History)* Committee later sent a letter to Lee Man Tat asking him to provide historical details concerning the founding of Lee Kum Kee, yet again to no avail. Subsequently, the leadership of Zhuhai CPPCC was said to have a specific discussion on how to win over his trust.

During the discussion, a Vice Chairman in charge of "sanbao" (referred broadly to three types of overseas Chinese beyond Mainland China)(2) liaison affairs mentioned that Choi May Ling's family originated from Beiling, Zhuhai, and she did maintain more regular contacts with her hometown and those institutions in charge of overseas Chinese affairs. They might therefore approach Lee Man Tat through Choi May Ling. The *Humanity and History* Committee thus visited Beiling to research into the background of the Choi family, and subsequently published the article of "Choi Hac Teng: The gentleman that I knew" in an issue of *Humanity and History*.(3) A few months later, Choi May Ling informed the *Humanity and History* Committee that Lee Man Tat had decided to donate to establish Lee Shiu Nan Memorial School in Nanshui. After that, Liang Zhenxing, Deputy Curator of Zhuhai City Museum and a member of Zhuhai CPPCC, personally paid a visit to Lee Man Tat and the Tai Po Headquarters of Lee Kum Kee in Hong Kong. He also conducted on-site investigations in Nanshui and Qibao, and subsequently published two historical accounts on Lee Kum Kee and the Lee family. In 1991, Lee Man Tat started to serve as Standing Committee Member of Zhuhai CPPCC (Situ & Tang, 1995; Lei, 2012: 45).

Lee Man Tat tried to gain an insight into the new development in Mainland China as early as the late 1970s, one example in this regard was a cruise along the Yangtze River in 1979. Soon after the Chinese Government decided to re-open its door to the outside world, it launched the first voyage for foreign ships to go upstream along Yangtze River to Central China in 1979. This first voyage was helmed by Tsoi Cheong Ming, by then the General Manager of The Hongkong and Yaumati Ferry Company. Tsoi Cheong Ming, a graduate of Wu Chang Maritime School, is also the brother-in-law of Lee Man Tat.(4) Lee Man Tat and Choi May Ling joined the trip that cruised along Yangtze River presumably not just for the scenery, but also to see for themselves the actual situation of Mainland China so as to make a better judgment on the feasibility of future investment.

According to Tsoi Cheong Ming, when the ship arrived in Wuhan, Lee Man Tat surprisingly spotted Lee Kum Kee Oyster Sauce on the shelves of the local Friendship Store. He was so excited that he bought out its stock and distributed one bottle to each of nearly a hundred group members. It is also noteworthy that the river cruise was highly regarded by the Chinese Government, as the then Deputy Director of Changjiang River Shipping Authority Gu Yonghuai accompanied the group all the way, acted as a tour guide along the Three Gorges to Gezhouba Dam, and briefed them about the contemporary social conditions and historical heritage of the locality.

While pondering over the locations of Lee Kum Kee's plants in Mainland China, the entrepreneurial Lee Man Tat also repeatedly attempted to break into its local market throughout the 1980s. His efforts started with a small-scale bid to test the response of its consumers and figure out their preferences. The earliest attempt should be selling Lee Kum Kee's products in the "Petty Trade Zone" in Zhuhai border. Goods traded in the Petty Trade Zone were tax-free, as opposed to a 70% tax imposed on imported goods in those days. Lee Kum Kee's products would be impossible to sell out under such heavy tax, given the much lower living standard of the locals. However, even though Lee Kum Kee could reach the locals tax-free in certain Petty Trade Zones, only its economical oyster sauce brand of "Choy Sun" was popular amongst the locals in the early 1980s, while the sales of other brands of its oyster sauce were lackluster.

In May 1984, the Rotating Palace Hotel had its sensational opening in Foshan. It was a joint venture of the "uncrowned Governor of Macau" Ho Yin, Ma Man Kei and Foshan City Government, with a revolving restaurant situated on its 18th floor, the very first of its kind in Guangdong Province which overlooked the old quarters of Foshan City. It was therefore regarded as the top business venue in Guangdong Province. Lee Man Tat instructed Chang Chin Vai to persuade the management of the revolving restaurant to sell Lee Kum Kee products, and subsequently sent them Lee Kum Kee oyster sauce for sale in the revolving restaurant. However, against expectation, it turned out that "after the first batch, there was no further order."

Lee Man Tat also decided to bring Lee Kum Kee back to Guangzhou after his earliest attempts of marketing in Zhuhai and Foshan. It had been a long lapse of over three decades since he left Guangzhou in 1949, and the city had changed beyond recognition. In other words, Lee Kum Kee had to start from scratch in this familiar yet unknown city. Lee Man Tat was fortunate enough to have the help of Shi Yushen — an alumnus of Choi May Ling from Lingnan University —who worked for Sun Yat-sen University in Guangzhou at the time.(5) Immediately after he had obtained the required business licence, Lee Man Tat set up a company at a room rented in Movie Star Hotel on Xin Gang Zhong Road, and began to sell oyster sauce in the local market.

Compared to the 1940s, Lee Kum Kee had no store in Guangzhou now, thus the sales of oyster sauce meant that they had to find the right wholesalers first, hoping to slip into Guangdong market via their local networks. Its Guangzhou office later moved to Huamei Building in Huadu District, and they even turned a rented room in Sun Yat-sen University's campus into its office for a while. Eddy Lee admitted that at that time, since the import tax was too high and they were short of capital to invest in local promotion, worse was the products' failure to match local tastes and the locals' consumption habits, Lee Man Tat's first battles on his return to the Mainland China market could be deemed as a complete failure.

From a closed state to opening-up, and from a planned economy to the experiment of market mechanisms, it is comprehensible that restrictions in an existing social system could not be shattered overnight. The same applies to the locals' consumption habits. Although Lee Man Tat was very keen to break into the potentially huge market of China and tried repeatedly, his early failure was conceivable given the various institutional constraints of the new market. As noted in classical theory of entrepreneurship, starting up new business and search for new opportunities could never be easy tasks. They entail a repeated exploration and toil for an insight to take shape, rather than simply popping up like a magician's trick. Likewise, Lee Man Tat was learning, while continued to search for a point of breakthrough.

Setting Up Factories in Fuzhou and Huangpu

Lee Man Tat could never forget his excitement when spotting Lee Kum Kee's oyster sauce on the shelves of Wuhan's Friendship Store and seeing the popularity of Choy Sun brand oyster sauce amongst the locals in Petty Trade Zone of Zhuhai. These experiences emboldened him to take a more aggressive approach in breaking into the Mainland market. After a series of trial and error, Lee Man Tat had also gained a more concrete hold of the socio-economic situation of Mainland China. Meanwhile, the embittered court case between Lee Man Lok and himself was wrapped up in 1986 as well. Indeed, he was short of cash due to the construction of Tai Po plant and his committed payment to Lee Man Lok, but he realised that it was time to start mapping out a clearer strategic layout for Mainland China.

Lee Man Tat set up a company in a rented room of the Movie Star Hotel in Guangzhou in 1992, and erected the signboard of Lee Kum Kee to promote its products.

In our interviews with Lee Man Tat, two major concerns came to light on his strategic layout for Mainland China. First of all, Lee Man Tat would like to ensure a stable supply of raw materials for Lee Kum Kee; secondly, he wanted to greatly upgrade Lee Kum Kee's production capacity by setting up new plants. As an ambitious entrepreneur with grand vision, Lee Man Tat had already planned ahead on the locations of new plants even before the completion of Tai Po plant in Hong Kong. It was not one but several plants in different geographical regions, with the oyster farming districts and other sources of raw materials also taken into account. Lee Man Tat had obviously learnt a lesson from the previous predicament that a newly-built Tin Wan plant could accommodate the company's need for a mere few years.

As we all know, Hong Kong underwent a period of industrialisation from the 1950s. It was a period when factories were everywhere and booming, and the rapid economic growth had made Hong Kong one of the "Four Little Dragons in Asia". In the meantime, Hong Kong's population was also expanding rapidly, and the subsequent rise in housing needs prompted a soaring land price. Lee Man Tat had long been ambitious to build a huge sauce kingdom, however, the exorbitant land price and rising wages in Hong Kong meant that Lee Kum Kee would never be able to find inexpensive industrial land for its new factories in this tiny territory. The only option was to go abroad for new plants. Consequently, its Los Angeles plant in the United States was completed in 1991, and that was followed by its Kuala Lumpur plant in Malaysia in 1997. Nevertheless, the only place that Lee Kum Kee could both build its plants cost-effectively and also enjoy the advantage of proximity was Mainland China.

The first place Lee Man Tat spotted to build a plant in Mainland China was Fuzhou. He had personally inspected the environment of Fuzhou in the 1980s, and was satisfied with its ample supply of seafood, such as its quality oysters and shrimps. Another factor that led to his selection of Fuzhou was the relationship with his friend Cheng Ah Ching, Principal of Hong Kong Mansfield College at that time. Lee Man Tat recalled that he knew Cheng Ah Ching very well and through him he came to know Chen Jingmiao,(6) who came from Fuzhou and was then the Vice Chairman of Hua Rong Limited. Hua Rong Limited was in fact the liaison office of Fuzhou City Government in Hong Kong, primarily missioned to draw investment from Hong Kong to facilitate Fuzhou's economic development.

When Lee Man Tat and Eddy Lee went to Fuzhou for a field study around 1988, it was Chen Jingmiao that guided them personally. Lee Man Tat soon decided to set up a seafood company in Fuzhou to procure and farm oysters as well as process raw materials. In the 1990s, he again invested in Fuzhou's Mawei High-tech Park to build a plant there. Eddy Lee pointed out that Fuzhou plant was in fact the place that produced Lee Kum Kee's first bottle of oyster sauce in Mainland China, and this was the economical brand of Choy Sun.

Lee Man Tat was so impatient that even before the construction of Mawei plant in Fuzhou, he had tried out other means to start the production. Immediately after Lee Kum Kee securing a production licence in Fuzhou, he rented a factory in Mawei for the production while building up the plant in Mawei High-tech Park. Xi Jinping was at that time Party Secretary of Fuzhou City, and Lee Kum Kee's Fuzhou plant was approved by him.(7) The plant immediately commenced production upon its completion in 1995, and that marked the starting point of Lee Kum Kee's investment in Mainland China.

It is worth noting that although Lee Man Tat is sociable and skilled at networking, he has always kept a low profile in Chinese business circle, and deliberately distanced himself from the political arena. He stayed for many years as CPPCC member at local level, and offered his advice to the authorities merely on fields that he was familiar with, such as the food and catering industry and family business development. His children have followed his low-key style in this regard, which can be seen as an established family ethos set by Lee Man Tat.

As mentioned, Fuzhou plant was just the beginning. Lee Man Tat continued to search for suitable sites in Huangpu, Guangzhou and Qibao, Xinhui later, ready for larger scale of investments (see discussion below). When both Huangpu plant and Xinhui plant were opened for production, Fuzhou plant ceased producing oyster sauce and was turned into a factory that processed raw materials. Today it focusses on processing oyster extracts and plums used for producing plum sauce. The plant was located in a remote suburb of Fuzhou, but since then the area has developed to be a bustling district with high-rise buildings and soaring property values.

Lee Kum Kee's second plant in Mainland China was located in the Huangpu sub-district of the Guangzhou Economic and Technological Development District, and its building process was more dramatic and challenging than the Fuzhou plant. Lee Kum Kee had made a deal with the authorities to build a plant there in 1988, which was the same year Tai Po plant in Hong Kong completed. Lee Kum Kee decided to buy a plot of around 200,000 square feet in the development district. It was well located with convenient road access around. However, the date set for payment and signing the sales agreement was 5 June 1989. It was indeed a day when China's political environment drastically changed and seriously impacted on the investment atmosphere of both Hong Kong and China.

Due to the sudden twist of the development and the unclear socio-economic prospects, Lee Man Tat and his sons were hesitant to sign the deal, which is understandable for any decision on long-term investment. Eddy Lee recalled that since all the formalities had been finalised, they eventually decided to honour the commitment to

Xi Jinping, the then Party Secretary of Fujian Province, was in discussion with Lee Man Tat, Eddy Lee, and Charlie Lee regarding their Fuzhou plant.

proceed and conclude the deal. That said, Lee Man Tat still felt that they needed more time to size up the situation. They therefore shelved the development plan of the plot temporarily.

From the summer 1989 to the spring of 1992, due to domestic impacts of the political turmoil, reformers within the Communist Party suffered a serious setback, while China was isolated by the United States and many other western countries. The process of reform and opening-up stalled and even regressed. It was only reactivated by Deng Xiaoping during his tour to southern China in Spring 1992 with a clear and powerful message to support the continuation of reform and opening-up. The octogenarian, albeit an ordinary party member unloaded all of his political and military duties when he toured the southern cities of Wuchang, Guangzhou, Shenzhen, Zhuhai, Shanghai, delivered enormously impactful speeches to reinforce the importance of economic construction. He boosted the reformers and reversed the tide, thus bringing China back to the track of reform and opening-up.(8)

In 1993, Lee Kum Kee was informed by Guangzhou Economic and Technological Development District(9) that since there was no construction on site, the plot was overdue, and the management committee had decided to reclaim the land right. It was indeed a breach of contract on Lee Kum Kee's side, and if the management committee insisted on reclaiming the land, according to the contract, it could forfeit the deposit. Lee Man Tat and Eddy Lee felt aggrieved because they did want to invest; they only delayed the process because of the uncertain political situation. Eddy Lee hurried to Guangzhou Economic and Technological Development District in an attempt to clarify the situation. Unfortunately, he failed to meet the officials concerned.

Fortunately, Lee Man Tat and Eddy Lee were well-acquainted with Zhao Yunhong,(10) who was President of the PLA First Military Medical University at the time. They got to know each other well because of their collaboration in Nanfang Lee Kum Kee (see discussion below in other sections). When Eddy Lee mentioned the problem to Zhao Yunhong, expressing their grievances as well as stressing their intention to proceed; unexpectedly, Zhao Yunhong offered to help. Eddy Lee recalled that:

> *He took out a brush and an envelope. I didn't know if there was anything in the envelope: I couldn't see it. He wrote four characters, and then another four, a total of only eight characters, and that's it. He slid the note into the envelope and passed it to me, saying that: "You hand this over to Guangzhou Economic and Technological Development District. Hopefully, they will help you." I therefore went back there again and delivered the letter. The problem was eventually solved.*

With the problem resolved, as in the case of the Fuzhou plant, even before the construction of Huangpu plant, the impatient Lee Man Tat rented a space in a nearby industrial building to set up a temporary factory, so that he could proceed with the production of oyster sauce and other products. It was not until 1998 that the Huangpu plant in Guangzhou was finally completed for full operation.

Compared to many Hong Kong industrialists who chose Shenzhen and Dongguan to build their plants in Mainland China in the 1980s, Lee Man Tat explained that although he had also inspected these places, he nevertheless settled the plants in Fuzhou and Guangzhou. Looking back, while their development paths were indeed littered with difficulties, problems were resolved one by one. The thorny road which Lee Kum Kee had gone through during this period of time was in fact a resonance of the tortuous path of China's reform and opening-up. An even more dazzling development awaited Lee Kum Kee on its way back to China, and that was its significant collaboration with the First Military Medical University.

From Nanfang Lee Kum Kee to Infinitus

By the early 1990s, Lee Kum Kee's main business remained oyster sauce. Shrimp paste and other condiments were less significant in terms of their contribution to revenue, and Lee Kum Kee was still unable to produce its own soy sauce. However, today, Lee Kum Kee has developed into two business arms, namely, the Sauce Group and the Health Products Group. On top of its century-old sauce business, the "Infinitus" health products business decisively founded by Lee Man Tat in 1992 is expanding in an amazing pace.

Infinitus was called "Nanfang Lee Kum Kee" before 2009. Its establishment is undoubtedly a classic case of constant entrepreneurship that embodied Lee Man Tat's spirit of "6677: proceed if you have 60% to 70% confidence to succeed" (see further discussion in Chapter 10). The road from Nanfang Lee Kum Kee to Infinitus was truly bumpy, and its prominent take-off since the 21st century was led by Sammy Lee. At its early stage of development, Nanfang Lee Kum Kee could barely survive a recurrence of catastrophic assaults between 1998 and 2000. It was Lee Man Tat's sturdy support at the critical moment that had stabilized the business, together with the toil and sweat of the core management team of Nanfang Lee Kum Kee.

Sammy Lee was the only child of Lee Man Tat that was not called back by him to serve Lee Kum Kee upon his graduation. He majored in financial management while studied at the University of Southern California, and as the youngest, he had such a leeway to work for Citibank Hong Kong for more than a year before joining Lee Kum Kee. Even after joining the family business, Sammy Lee went through a comparatively different course of development. He first handled the company's finance. However, as entrepreneurial as his father and always enjoyed starting up new businesses, he tried out catering business and real estate investment under the auspices of the Group

An aerial view of Lee Kum Kee's production base in Huangpu, Guangzhou.

at various times. In 1991, Sammy Lee set up Hong Lai Dou Limited with his father's support and developed a restaurant chain called "Kin Yat Kitchens" which offered reasonably priced Chinese-style home-cooking dishes in housing estates and residential communities. Within five years, Kin Yat Kitchens expanded to 12 restaurants across Hong Kong (Lee, 2012: 198-199). Apart from Kin Yat Kitchens, Sammy Lee was also involved in another important business in 1992, which was Nanfang Lee Kum Kee.

In retrospect, the business of Nanfang Lee Kum Kee appeared quite accidentally. Lee Man Tat's ability to see the big scene from the small one again prompted him to grasp the opportunity timely. As the saying goes: "Opportunity awaits those who are prepared." Lee Man Tat had persistently been looking for opportunities in Mainland China and was well-prepared, thus when they appeared, he grasped them tight and never let go against difficulties. But how did such an opportunity emerge in the first place? The story should begin from Poon Kwong Yu, a former classmate of Sammy Lee in St. Joseph's Primary School.

According to Lee Man Tat, Eddy Lee and Sammy Lee, Poon Kwong Yu's mother used to collaborate with the PLA First Military Medical University(11) in Guangzhou for business. She was entrusted to introduce or arrange patients from Hong Kong to a private hospital called "Huiqiao Lou" for medical check-up and treatment. Huiqiao Lou was run by the First Military Medical University. It was a profitable business as medical services in Mainland China were relatively cheap and competitive in those days and certain aspects of its medical technology were uniquely advanced.

As the private hospital's Chinese name (Huiqiao Lou) literally meant a hospital that "benefits overseas Chinese", it mainly served non-local Chinese from Taiwan, Hong Kong, Macau and Overseas communities. It was part of the measures of "earning foreign currencies and

Being a gastronome, Lee Man Tat (2nd left) not only participated in the running of the Sha Tin Floating Restaurant in the 1960s, but also supported his son Sammy Lee (1st right) to found the "Kin Yat Kitchens" years later. These efforts were examples of his constant entrepreneurial spirit.

improving revenue" taken by the university in the early days of reform and opening up. Its successful operation proved that Zhao Yunhong, the then President of the university, was an entrepreneur apart from his leading role in medical education. Mrs. Poon once chatted with Lee Man Tat about the private hospital, and casually asked Lee Man Tat if he had any interest to collaborate with the university. Lee Man Tat astutely sensed the great potential of a business in the field and immediately sought a meeting with Professor Zhao.

Lee Man Tat, together with his sons, met Zhao Yunhong in Guangzhou on 8 August 1992. Zhao Yunhong right away brought up in their first meeting that the university had a research output of compounded polysaccharides which was highly effective clinically as a healthcare supplement, and that they wanted to test the market. Lee Man Tat listened attentively. He had long thought about the business of healthcare products based on Chinese herbal medicine, but he confessed that he knew nothing about these chemical and pharmaceutical jargons. He observed that: "President Zhao put forward the names of their research outputs, which I totally had no idea. I thought these were promising, thus we shook hands in less than 60 minutes and confirmed our intention to collaborate. That's the way the business began, and today, its development is far beyond my expectation."

Eddy Lee also confessed that: "Their research outputs and compounded polysaccharides were all out of my depth. I really didn't know what they were. We had a good discussion anyway, and I remember that we shook hands for collaboration in the car and agreed that the university accounted for 45% of the shares, while Lee Kum Kee the remaining 55%. Although the university was entitled to 45% of the shares, no capital investment was needed. They were responsible for the products' know-how and the provision of a plot in its campus to build the plant." Eddy Lee added that the university later also turned

Lee Man Tat and his four sons with former Governor of Guangdong
Ye Xuanping (4th right) and former President of the First Military
Medical University Zhao Yunhong (middle) in Guangzhou. All were
looking forward to joining hands in their new project.

one of its campus dormitories into the provisional office of Nanfang Lee Kum Kee.

From Lee Man Tat's various preparatory works in the early 1990s, it was obvious that he did not simply follow Zhao Yunhong's proposition and rush into the healthcare industry of Chinese herbal medicine in the way that he described. Clearly, he had long borne an idea and vision to start a business in this field and, to him, it was a long-term prospect for at least several decades. Harry Yeung, Senior Vice President of Lee Kum Kee Health Products Group, mentioned that when the company headhunted him from the British-owned Glaxo Hong Kong Limited in 1994, he was already the General Manager of Glaxo Hong Kong. He was entrusted with the mission to develop Lee Kum Kee's new business in Chinese herbal medicine. Lee Man Tat shared with him his ambition to invest even as long as a century to develop the know-how in the field, even though Lee Kum Kee was a new comer then. Lee Man Tat realised from the inception that the Chinese herbal medicine and Chinese health regimen would have boundless business opportunities in a progressively wealthier society, in which people are paying more attention to their health as well as to environmental protection. Yet, it was a market fairly undeveloped.

In 1992, Nanfang Lee Kum Kee Health Products Co. Ltd. was established. The company must build its factory for the production first, thus Lee Man Tat entrusted Eddy Lee for the building of its factory and Sammy Lee for sales. The factory was situated in the campus of the university, and for that reason, Eddy Lee frequently visited the university during those days. In 1994, the factory was completed and production began, however, the first-year sales of their product, Health Tonic, were very poor. Eddy Lee noticed that "if we put up advertisement, we could sell out some; otherwise, we could hardly move the market." Given the poor turnover, for a while, they had to let the factory workers go home and wait for call to work.

In the face of great difficulty, Sammy Lee, as the commander on the frontline, was anxious to find a way out and thought hard. It finally dawned on him that healthcare products are by nature different from other commodities and are difficult for the consumers to feel or see their immediate effects. Healthcare products must therefore rely on the interaction of people to establish their rapport and gradually build up their reputations by word of mouth. In short, their sales could not simply be based on advertising hype. Sammy Lee subsequently returned to Guangzhou to develop a direct-sales system and finally opened up the market. Since he managed to find the right way for sales, Nanfang Lee Kum Kee soon entered a rapid stage of development.

Return to Qibao, Xinhui

Lee Man Tat was already 66 years old in 1995, but he was still strong and ambitious, with a fiery spirit of constant entrepreneurship. It was in that year that he made probably the most courageous decision in his life: back to Qibao, Xinhui and invested heavily in his ancestral hometown. This momentous choice not only profoundly affected the course of Lee Kum Kee's future development, but the trajectories of its subsidiaries as well, including Infinitus. Lee Man Tat's strong affections to its ancestral hometown could also be seen in this process.

Lee Man Tat's investments in Mawei, Fuzhou and Huangpu, Guangzhou during the 1990s were far from the conclusion of his efforts to build his sauce kingdom. In fact, since his return trip to Mainland China in the late 1970s, Lee Man Tat had thought about the construction of a larger and more permanent production base in China. His sauce kingdom included not only oyster sauce and shrimp paste, but the more widely consumed soy sauce. However, a production base for soy sauce entailed a much larger space for the plant and a heavier investment in facilities than those for oyster sauce and other condiments. Furthermore, other conditions of production, such as the intensity of sunshine and water quality, were also elements that he needed to take into account.

Lee Man Tat recalled that in the 1990s, Lee Kum Kee was still unable to produce soy sauce, and that was haunting him all the time. To this end, he organised a special team to study the know-how in producing soy sauce, and even bought a traditional soy sauce factory in Kwu Tung, Sheung Shui as research base for the special team (see Chapter 8 for a detailed account). But Lee Man Tat would not wait until the moment of his breakthrough to search for a site for such factory.

Where should this factory be placed? Around the Pearl River Delta, Lee Man Tat and Eddy Lee had been searching for more than a year, travelling through Zhongshan, Foshan, Huizhou, Nanshui (in Zhuhai City), Pinghu (in Shenzhen City), Conghua (in Guangzhou City), Dongguan and others. Yet, they were still unable to make a final decision. Each location had its limitations. For instance, they did not choose Pinghu, even though it was a location very close to Hong Kong and the local government was also generous enough to offer a vast 10 million square feet of land, yet it lacked water resource. As for Nanshui, the local government offered the best deal, relatively speaking, with a 400 mu of land plus an exclusive seashore section with high-quality seawater for Lee Kum Kee to farm its oysters. However, it fell through because the then Party Secretary of Zhuhai City, Liang Guangda, required a lump sum payment for the investment, while Lee Man Tat was after payment over three instalments.

Lee Man Tat finally selected Qibao, Xinhui not only because the offers of the local authorities were attractive, but also due to his hometown nostalgia. As described at the outset of this book, Lee Man Tat was not persuaded by the local cadres on his first trip to Qibao in October 1990. That night, the locals' passion was so overwhelmed and touching that he was really moved. However, the traffic barrier of "seven rounds of

Bottles of Lee Kum Kee Oyster Sauce on the production line with a worker inspecting the quality.

river crossing via rafts" was too formidable to be overcome. But the problem was largely resolved in December 1992 when the Wong Haking Bridge was opened for traffic. The bridge also rekindled hopes of Qibao's local cadres in luring Lee Man Tat's investment to his ancestral hometown. In the following two or three years, they kept visiting Lee Man Tat in Hong Kong during the Chinese New Year, briefing him about the latest developments of Qibao, and repeatedly lobbying for his investment.

Apart from the bridge, they had good offer of land for the investment. Wong Haking Bridge was completed in December 1992, and by that time, Deng Xiaoping had wrapped up his tour to southern China in early 1992 and empowered the flexibility of investment policy to the local cadres. Li Guoli,(12) the then Deputy Town Mayor of Qibao, commented that exactly because the local cadres were "less familiar and stringent with the regulations than Zhuhai's cadres," they were bolder to offer more land for foreign investors, and the immediate commitment was a 200 mu of land for Lee Kum Kee.

However, to their surprise, Lee Man Tat responded that 200 mu were simply too little, he wanted 500 mu. In his mind was not a small plant, but a long-term production base for his sauce kingdom. Li Guoli found it prickly, as local cadres of Qibao were not of sufficiently high ranking in government administration to approve such a large plot for foreign investors. Moreover, their superiors might subsequently challenge the necessity of such a large plot for a sauce company and might suspect that there could be collusion or even corruption in the process for personal gains.

Lee Man Tat's pledge for at least 500 mu of land not only shocked the local cadres of Qibao, but his children as well, all questioning that it might not be a sound judgment. Lee Man Tat later kept asking for more land in Qibao for the expansion of the production base, and that deeply worried Elizabeth Mok, who thought to herself: "Would the company collapse? But daddy never seemed worried." Lee Man Tat confessed in interview that it was indeed risky, but he was glad that he was still in full charge of the company, thus able to impose his will: "I could do whatever I decided to do. I said I wanted to buy that huge piece of land in Xinhui, and I bought it. In fact, we still didn't know how to produce soy sauce at that time, not until 2000."

Lee Man Tat's request to increase the allocation of land accorded with the hope of the local cadres to encourage investments to revitalise the local economy. Under such circumstances, they finally found the way round to resolve the problems and approved the land allocation.(13) Nevertheless, when the construction of the plant started in 1995, the local cadres found out that Lee Kum Kee only built a small factory at a corner of the plot, and left most of the land untouched. They became nervous and urged for the speeding up of their development. Eddy Lee revealed that in order to avoid unnecessary problems, Lee Man Tat deliberately erected buildings at both ends of the plot first, and left the central part for future development lest the authority became suspicious of land speculation.

The huge project of Qibao plant started in March 1995. The first phase of the construction plan was completed in April 1996 for the plant to begin production. The factory was quite primitive and lacked most facilities to cater for employees' needs, thus making their lives truly difficult. Yang Jieming arrived there in early November 1996 as a pioneer for soy sauce production. She recalled that piling works of some buildings were still being done upon her arrival:

> At that time, it was only a plot of 200 mu and no dormitory at all. The reeds grew to people's armpits. Right in front of the road you could find only one small building, which was the police post responsible for looking after the bridge. Initially, we had only six or seven people, and

I was the only female. After nightfall, the whole road was really dark with no light at all. Qibao Bridge would merely light up once a year, which was the first night of the Chinese New Year. Since it was a wide and open space, once it was hit by lightning, no matter where it hit, we would suffer a complete blackout. Transportation was also a big issue here, very inconvenient indeed.

After several years of pioneering efforts, Lee Man Tat finally led the special team to achieve a technological breakthrough in soy sauce production. The company subsequently further expanded its Qibao plant in Xinhui. Within five years, the plant had expanded from less than 200 mu to 1,700 mu; and it has kept expanding to 2,000 mu today. This is a clear and solid indication of Lee Kum Kee's speedy pace of growth.

Conclusion

China's reform and opening-up policy since the late 1970s finally ushered in a period of peaceful development, as well as a golden opportunity for the rejuvenation of the nation. Lee Man Tat had entrenched Lee Kum Kee's sales network in the overseas Chinese communities worldwide by the 1980s, and he sharply sensed the potential of the China market. After much deliberation, he decided to open the market up with full force, and invested heavily and repeatedly. Lee Man Tat's solid investment in China has supported its reform and opening-up. And although it was an occasionally stormy course, his investment has also beefed up his ancestral hometown. Looking back, it was in no doubt a win-win situation to Lee Man Tat himself, his family business Lee Kum Kee, and the motherland of China.

Today, Lee Kum Kee's plant in Qibao, Xinhui is conspicuous by numerous rows of giant tanks for fermentation of soybeans through the natural and traditional South-China technology. The plant has also become a landmark of Qibao to the pride of the locals. To Lee Kum Kee, Qibao plant is currently its largest production base, particularly concerning soy sauce and its associated products. Similarly, to Qibao, Lee Kum Kee Sauce Group and Lee Kum Kee Health Products Group constitute its largest enterprise as well as the prime employer and tax payer. In other words, Lee Kum Kee is its main source of revenue. The two are simply inseparable today. Lee Man Tat has repeatedly impressed upon his children that he had selected Qibao to keep ties between Qibao and his family: "I want you all to come back here regularly…" The message behind these words is clear: it's not just about business, but the strong affections of the family towards its ancestral homeland.

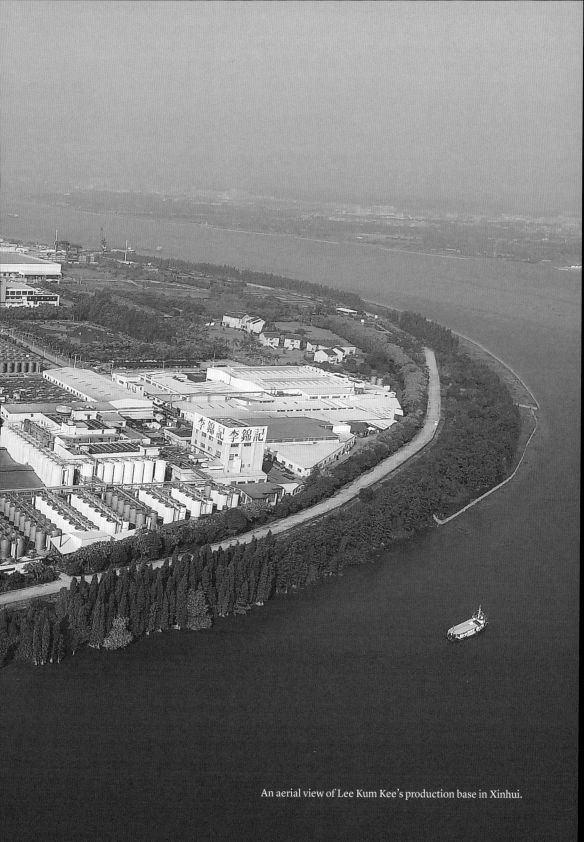

An aerial view of Lee Kum Kee's production base in Xinhui.

1   There are many statistics about the overall number of overseas Chinese. It is roughly estimated between 30 million to nearly 90 million (Zhuang, 2009). A more systematic and authoritative statistics pointed out that as of 2011, there were 40.3 million overseas Chinese in 148 countries, and almost half of them settled in Southeast Asia (Poston & Wong, 2016: 362).

2   It referred to the "three types of compatriots," namely the compatriots from Hong Kong and Macau, the compatriots from Taiwan, and the compatriots from overseas communities.

3   The article, written by Wen Dingming, was re-published in *Humanity and History of Zhuhai (Issue No.9)* in December 1990. Liang Zhenxing's "Lee Kum Sheung and the Lee Kum Kee Oyster Sauce" was also re-published in the same issue (Wen, 1990; Liang, 1990).

4   Tsoi Cheong Ming joined Lee Kum Kee as well later. He was appointed General Manager of Lee Kum Kee's Los Angeles office in 1986, responsible for the United States market until his retirement.

5   Shi Yushen was Secretary General of the Alumni Association of Lingnan University in the 1980s.

6   Chen Jingmiao stationed in Hong Kong from 1985 as the Vice Chairman of Hua Rong Limited, and was transferred back to Mainland China in 1992. He was once a teacher of Fuzhou No.3 Middle School, and became Deputy Mayor of Fuzhou City before stationing in Hong Kong.

7   Xi Jinping served in that capacity from 1990 to 1996.

8   Deng Xiaoping's tour to southern China was taken between January 18 and February 21 in 1992, which lasted for nearly one month. The first day of the Chinese New Year in 1992 fell on February 4.

9   Apart from the post, Huang Longyun concurrently served as a Party Standing Committee Member of Guangzhou City from 1993 to 1995. He was ranked the same level as Deputy Mayor of Guangzhou City at the time.

10   Zhao Yunhong was born in August 1922 in Nantong City, Jiangsu Province. He joined the New Fourth Army in 1939 and served as Military Surgeon, Health Captain, Superintendent of Base Hospital and Column Health Minister during the Anti-Japanese War. He assumed Presidency of the First Military Medical University in 1975 and served the university for over a decade. He

retired from the post at the rank of Major General, and died on 9 November 2002 (*Nanfang Ribao*, 19 November 2002).

11   The PLA First Military Medical University was transferred from the military to Guangdong Provincial Government in August 2004, and subsequently renamed "Southern Medical University".

12   Li Guoli is former Deputy Town Mayor, former Town Mayor and former Party Secretary of Qibao Town. After his retirement, he joined Lee Kum Kee and served as External Affairs Director. He is currently Counsellor for External Affairs.

13   This is based on our interviews with retired local cadres previously involved in the case. They were He Xiansong, Li Zhinong, Li Jiuru, Li Quansheng and Li Baihan. He Xiansong is former Mayor and former Party Secretary of Xinhui City; he is now retired. Li Zhinong once served as Deputy Town Mayor and Town Mayor of Qibao Town, and is currently Chief of Transportation Bureau, Xinhui District. Li Jiuru is former Director of the Overseas Chinese Affairs Office, Qibao Town and former Vice Chairman of the Federation of Returned Overseas Chinese, Huicheng Town; he is now retired. Li Quansheng is former Town Mayor and former Party Secretary of Huicheng Town; he is now retired. Li Baihan once served as Head of the Third Production Team, Chongli Village, Qibao Town, and is currently Director of Tanchong Management Area.

Lee Man Tat was inducted into the Babson College's "Academy of Distinguished Entrepreneurs" in the United States in 2008: recognition by one of the world's top academic institutions for his outstanding achievements.

GO
WORLDWIDE
IN CRISES

If any of you bears no confidence to continue in the business,
you can get up and leave now... I have absolute confidence in the
company's future, and I will give you my full support. We will
definitely overcome difficulties and get to a better development.

Lee Man Tat

INTRODUCTION

When considering strategy for Lee Kum Kee's development, Lee Man Tat had never focused his attention on merely one point, one way, or one dimension, but seeing the big picture from the small detail, and dealt with it comprehensively from different angles based on different scenarios. We have mentioned in the previous chapter that by the mid-1980s, when Lee Man Tat had settled his family crisis and turned his full attention back to the business, he began to explore opportunities in different parts of Mainland China. He did it on multiple fronts at the same time: for deals and plans which were ready to go, he took immediate actions; whereas for those not yet ready, he kept working his way through. In short, Lee Man Tat was ardently devoted to building up his sauce kingdom in the near future.

By the 1990s, Lee Man Tat had successfully established several production bases in Mainland China, and began to make inroads into its market. Meanwhile, he was clearly aware of the overwhelming trend of globalisation, and adjusted his goal accordingly: from then on, he strived to forge Lee Kum Kee as a truly global brand which goes beyond the overseas Chinese communities, and is able to rival other world-class sauce brands. Such actions and efforts not only brought Lee Kum Kee back to China but worldwide, and amazingly, even beyond the limit of the sky in voyages to the outer space.

Making Soy Sauce, Finally

When Lee Kum Kee's Xinhui plant in Guangdong Province commenced construction in 1995, Lee Man Tat had not yet mastered the technology of producing soy sauce. He and his special team were still learning and experimenting, through trial and error. It was in fact risky that the Xinhui plant based a great deal of its design in catering to the production of soy sauce and its derivative products. Commanding the know-how of making soy sauce to further expanding his sauce kingdom was indeed a strong aspiration of Lee Man Tat for decades.

Soon after Lee Man Tat took helm of Lee Kum Kee in 1972, he was clearly aware that apart from oyster sauce, the company should also produce other condiments. Although a single product could enjoy the advantage of specialisation and a more effective operational management, the fact was Lee Kum Kee's oyster sauce had long been the leader in the industry and commanded a lion's share of the market. If it continued to produce oyster sauce alone, it would be hard for Lee Kum Kee to achieve breakthrough and grow stronger. And last but not least, if the overall market circumstances or the preference of mainstream customers changed, the only key product of a company might easily be eliminated, and hence seriously cripple the company.

Lee Man Tat astutely saw the potential of soy sauce as the most demanded Chinese condiment in the food industry. As a man with outstanding business acumen and strong desire to excel, making soy sauce was like his sacred mission to build his sauce kingdom. Moreover, soy sauce is an indispensable ingredient of a variety of Lee Kum Kee's sauces, and the soy residue from soy sauce production can also be used as by-products. Therefore, even from the perspective of cost reduction, Lee Kum Kee must be able to produce its own soy sauce.

However, making soy sauce was far from an easy task, even to a sizable sauce enterprise like Lee Kum Kee. It was a tortuous process full of setbacks, which had entailed the company paying substantially for failures. Lee Man Tat's perseverance finally prevailed. Back to the early days in the 1980s, Lee Man Tat began his attempt to enter the soy sauce market. With little expertise on the production process, he resorted to acquiring soy sauce from reliable manufacturers, and then packaged it for sale under the brand of Lee Kum Kee. But in one shipment to America during the 1980s, Lee Kum Kee suffered a loss of over 30 containers of its soy sauce due to quality problem.

Lee Man Tat was informed that the specific batch of soy sauce had deteriorated. In order to verify the complaint, "we opened the bottles to check, and it was true. We had to pay them compensation." The event plagued Lee Man Tat for years thereafter. He was upset about the loss particularly at a time when money was tight because of the company's heavy investment in building Tai Po plant and his break-up with Lee Man Lok: the loss was a hefty blow to the company's finance. But the intangible blow of the event to Lee Kum Kee's renowned reputation was even more heartbreaking to him. Lee Man Tat indeed collaborated with credible soy sauce manufacturers, but he had no control on their production process and storage, and that unfortunately had led to the damaging event. As he always regards quality as the life-blood of Lee Kum Kee, the incident was unbearable to him, and hence reinforced his determination to make his own soy sauce.

The process of making soy sauce from scratch tested Lee Man Tat's wisdom in matching the right people to the right tasks, and once again, revealed his relentless pursuit of quality. First and foremost was to find professional yet trustworthy experts for help. The two key figures, in this regard, were Professor Yuan Zhenyuan from the South China University of Technology and Yang Jieming, another expert recommended by the former.

Lee Man Tat always makes good use of his interpersonal network. In this case, Choi May Ling's network amongst the alumni of Lingnan University was of great help. It was exactly through this network that Lee Man Tat got to know Professor Yuan Zhenyuan, an alumnus of Lingnan University. Yuan Zhenyuan was a well-known figure in the Mainland condiment industry, and was familiar with the "ins and outs" of the entire industry. Like many intellectuals, he was repeatedly persecuted and disparaged as a "stinking old ninth" (a Chinese dysphemism for intellectuals) during the Cultural Revolution, and was sent to work in a soy sauce factory at the grassroots level. Through such experience, Yuan Zhenyuan became very familiar with the flow of production of the industry, right from the most basic operations. Moreover, he knew prominent and influential figures in the industry pretty well. Lee Man Tat entrusted Yuan Zhenyuan to find somebody in the field for help, and Yuan Zhenyuan recommended Yang Jieming, a lady who was equally familiar with the industry.

In 1993, upon Lee Man Tat's invitation, Yang Jieming started to assist him in designing the soy sauce plant as an advisor. At that time, Yang Jieming was still an employee of a research institute, thus she could only offer her service indirectly. Still, Lee Man Tat and his sons took it seriously and separately talked to her. "They were meticulous in judging and making decision; trying to know me better from different aspects." In order to form a professional team, Yang Jieming later approached two other experts from the South China University of Technology: one was a professor specialised in machinery and another was a lecturer from the university's School of Light Industry and Food Science. Together, they planned the blueprint of the plant for three years, including its detailed design and budget, and were informed by the company in 1996 that it had selected a site in Xinhui, Guangdong for the plant. It was also in 1996 that Yang Jieming officially joined Lee Kum Kee, and became the key technical staff for its production base in Xinhui. She is today the Chief Technologist of Lee Kum Kee Sauce

Group.

Yang Jieming noted that Lee Kum Kee was highly meticulous in its preparation for building the new plant in Xinhui. The management team scrutinised each proposal exhaustively, aiming to achieve the best. The same spirit was also reflected in the learning process of making soy sauce led by Lee Man Tat. In fact, before the construction of Xinhui plant, Lee Man Tat had bought a local small sauce factory (Chue Lung Factory) in Kwu Tung, Sheung Shui, Hong Kong, and turned it into a secret base to experiment soy sauce production. He invited Yuan Zhenyuan from Guangzhou to station there for a few months, leading the special team to make soy sauce. They used the shabby buildings and drying yard of this secluded sauce factory to experiment the production process of soy sauce, without any professional facilities. First, they used the only stove and the only big wok to boil soybeans, cooled them down afterwards, and sent them into the Koji room. The next step was to let them ferment in the fermentation pool. The gist of this experiment was to try to adopt the most authentic and traditional South-China way in making soy sauce, in the hope that it could really work.

Apart from experimenting the production process, team members in Kwu Tung also needed to learn how to cultivate and nurture strains, so as to maintain the quality and stability of soy sauce. Lee Man Tat once shared with Yang Jieming the soy sauce blunder that he had made in the 1980s, stressing that, "We don't want to produce lousy products. The quality of my soy sauce must be of good quality." Tony Mok, General Manager of the Xinhui and Huangpu plants, was one of the team members in Kwu Tung when he was still a junior. He also noticed that Lee Man Tat and all his children consistently attached paramount importance to quality during the course of studying soy sauce production, and invested heavily in terms of money and time without any hesitation.[1]

The Chue Lung Factory in Kwu Tung, Sheung Shui once served as a research base of Lee Kum Kee in making soy sauce. Few would realise that behind the shabby façade of the factory, a series of innovative experiments leading to a major accomplishment was taking place.

When the team had finally succeeded in their experiment in Kwu Tung, Lee Man Tat applied the same flow of soy sauce production on Xinhui plant. Nevertheless, for the sake of prudence, the team again conducted small-scale production test in Xinhui, starting from a mere 2,000 tonnes. Yang Jieming clearly remembered that at the beginning, they only ordered 38 sets of 30-cubic-metre fermentation tank. Later, in phases, they upgraded these tanks to 60-cubic-metre and again to 120-cubic-metre today, indicating the growing productivity of Xinhui plant. Material used to make these tanks was also a serious issue cautiously studied by the team. The fibreglass tanks adopted in Xinhui plant were never before used by other soy sauce manufacturers in Guangdong Province. Stainless steel was the best, yet very expensive. After consulting experts from the South China University of Technology, Yang Jieming proposed to use fibreglass, which was a material fell in between carbon steel and stainless steel.(2)

With approval by Lee Man Tat and Charlie Lee, the team searched all over China for the best fibreglass manufacturer and to ensure that fermentation tanks concerned were in line with food safety standards. While Yang Jieming conducted research to solve technical problems, Lee Man Tat was attentive throughout the process. Though he did not really understand the technical details, Yang Jieming could always expect his support for her bold attempts. And more importantly, when procuring related facilities, "we would list out options for each facility, I mean what was the grade concerned and how much it would cost; and let him choose. He would always choose the best one."

A new milestone of Lee Kum Kee in its soy sauce production was marked in November 2013 when the company purchased the patented "Continuous Steam Cooking and Rotary Type Koji Making Machine" from Japan's Fujiwara Techno-Art Company Ltd. After nearly two years of joint research to improve its functions and work on its installation, the facilities were officially inaugurated in October 2015. It is the most advanced soy sauce production system in the world, notably featured by its dual-capability to manufacture the Cantonese-style soy sauce using fermented soybeans and the Japanese-style soy sauce using soybean derivatives. The new soy sauce production plant could also greatly enhance soy sauce productivity as well as stability of its quality. It is equipped with fully automated facilities for washing, soaking, pre-heating and continuous steam cooking the soybeans, as well as wheat flour mixing, inoculation, material input, koji making, koji output, washing and sterilization. With its annual soy sauce productivity of 180,000 tonnes, Xinhui plant today can produce a total of 500,000 tonnes of soy sauce annually.

The installation of such advanced facilities not only fulfilled Lee Man Tat's long-cherished aspiration, but also realised the common dream of the two family businesses, Lee Kum Kee and Fujiwara, in joining hands to build a world-class soy sauce production system since their first encounter two decades ago. Xinhui plant commenced its soy sauce production in small-scale in August 1998. Akio Fujiwara, the late President of Fujiwara, learnt about it and paid a visit to Lee Man Tat together with his wife Sadako Fujiwara. He was eager to introduce such purposely-designed facilities to Lee Kum Kee. However, at that time, the annual soy sauce productivity of Xinhui plant was a mere 2,000 tonnes, thus not justifying such a heavy investment. Akio Fujiwara failed to make the deal, but friendship of the two families sustained, and Lee Man Tat never stopped yearning for this dream facilities.

When the time was ripe in 2013, Lee Man Tat decided to purchase the facilities. However, Akio Fujiwara had already passed away, and it was the third generation of Fujiwara — Akio's eldest daughter Keiko Fujiwara, who signed the agreement on behalf of the company. Also participated in the signing ceremony was Akio's widow, Sadako Fujiwara, who held Akio's portrait in her hands during the ceremony. As in the case of cross-generational friendship between the Lee family and the Chen family of Japan's Daiei Trading Company, Lee Man Tat's children have now also established a strong bond with the current generation of Fujiwara. It is again the continuation of a practice cherished by Lee Man Tat, in which a sincere human relationship goes far beyond business partnership.

Lee Man Tat set the ambitious goal of soy sauce production for Lee Kum Kee on his way to building his sauce kingdom. He paid heavily for the lesson learnt from the blunder of loss due to soy sauce deterioration beyond his control. Since then, he had been spending a great deal of resources for years to learn the know-how of soy sauce production until his perseverance prevailed, eventually. It was again a great leap forward to glory for Lee Kum Kee.

The Crises of 1998

Although Lee Man Tat had achieved outstanding breakthroughs in various aspects in the 1990s, Lee Kum Kee's advancement was threatened by uncertainties. The crises of 1998 had their roots primarily in the still transforming socio-political circumstances in Mainland China, but the subsequent conflict of views amongst family members was a trickier issue. In short, the crises were multi-layer: first of all, Lee Kum Kee Group's overall financial stability was at stake; secondly, the viability of Nanfang Lee Kum Kee became questionable; and thirdly, Sammy Lee's aspiration of leaving the Group to stand on his own feet. For Lee Man Tat who consistently emphasised that "family is above the enterprise," the attempt of any family member to leave the Group constituted the gravest crisis of all.

Let us first review the overall financial stability of Lee Kum Kee Group in the 1990s. Lee Kum Kee began to consider a new strategy of business diversification since 1990, particularly in meeting the challenges of globalisation while evolving itself into a worldwide brand. Sammy Lee's

Photo taken on Akio Fujiwara's visit to Lee Kum Kee. It was the beginning of a sincere friendship between the late President of Fujiwara Techno-Art Company and Lee Man Tat.

Kin Yat Kitchens was the first successful attempt to follow such strategy in 1991. The Group then set up Lee Kum Kee Properties Ltd. in 1992, dabbling in the field of real estate investment which Lee Man Tat had himself involved for long. Two years later, in 1994, Charlie Lee led the establishment of LKK Trans Ltd. to exploit transport business between Hong Kong and Mainland China.

The Group again established Vac Vem (Hong Kong) Packaging Company Ltd. in 1995, which specialised in the production of high-quality metal caps to synergise with its growing soy sauce business. In 1996, the new Caring International Ltd. was formed to aim at marketing health and personal-care products to a rising affluent middle class. It was also in 1996 that Lee Kum Kee collaborated with Hong Kong University of Science and Technology to set up the "Hong Kong Traditional Chinese Medicine Research Centre", bearing a grand vision to promote the modernisation of traditional Chinese medicine.

When the region was suddenly hit by the Asian Financial Crisis between 1997 and 1998, Lee Kum Kee was caught off guard. The approach of diversification in its business and the subsequent expansion of its investment had led to an increase of financing cost during the crisis, causing cash flow problems. Charlie Lee recalled that Lee Kum Kee's real estate investments were particularly hard hit by the slump of housing market. Given the severe challenge ahead, according to David Lee and Ho O Shang of the Finance Department in Lee Kum Kee Group, Lee Man Tat decided to re-focus on the core business of sauce and condiments after much deliberation with his management team. The conservative approach meant that the Group would wind up non-core business one by one, including the profitable Kin Yat Kitchens founded by Sammy Lee. Lee Kum Kee Group regained its financial stability soon after it cleared itself from non-core business, which was then no longer an issue.

Lee Man Tat decided to purchase facilities from Fujiwara Techno-Art Company in 2013 to be installed at Lee Kum Kee's production base in Xinhui. Although Akio Fujiwara had already passed away, his widow and daughter attended the signing ceremony with his portrait in centre place. It was indeed a very moving scene. Charlie Lee expressed gratitude to Mrs. Fujiwara on behalf of Lee Man Tat.

In comparison, the situation of Nanfang Lee Kum Kee in 1998 was much worse. Lee Man Tat seldom meddled in the business of Nanfang Lee Kum Kee as he was not familiar with health products. It was therefore the baby of Sammy Lee primarily run by him from the inception. Nanfang Lee Kum Kee was on track and became increasingly thriving after Sammy Lee had changed its business model from retail to direct sales, yet unexpectedly, the company had been seriously affected by a new policy adopted by the Central Government. The State Council, on the basis of various considerations, suddenly announced a strict policy on 21 April 1998 to ban any form of direct selling. Nanfang Lee Kum Kee as a direct sales company immediately found itself landed in a predicament with almost no way out. Its sales slumped and inventory piled up while pending transition, but nobody had a clear idea of where to go under such uncertainty.

It never rains but it pours. In late 1998, the State Council introduced another strict policy of banning the military and all its associated units from engaging in commercial activities. The PLA First Military Medical University, under such policy, was forced to abruptly end its collaboration with Nanfang Lee Kum Kee. This second blow to Nanfang Lee Kum Kee in 1998 was equally heavy, if not heavier. Lee Man Tat purchased the remaining shares of the company from the university, and subsequently made it a wholly-owned company under Lee Kum Kee Group. However, its viability at the time became questionable.

Lam Yu,(3) Global CEO of Infinitus, is a former military surgeon once served the First Military Medical University. He recalled in the interview that after consecutively struck by the two drastic policy changes in 1998, Nanfang Lee Kum Kee suffered a 90% drop in its turnover, and remained in such dire condition for over a year. Pessimistic feeling was pervasive amongst its staff, and the bleak prospect prompted them to fear of the company winding up.

Sammy Lee recalled that since the State Council remained unclear in its strategic direction, Nanfang Lee Kum Kee continuously suffered from great financial losses, and that inevitably affected the financial performance of its parent company Lee Kum Kee. As the supreme commander of the Group, Lee Man Tat was in line with the Group's mainstream opinion at the outset, which was to sell or wind up Nanfang Lee Kum Kee. The four sons of Lee Man Tat, however, could not reach a consensus on its fate. During the debate, Sammy Lee began to moot the idea of bringing his baby out of Lee Kum Kee and to stand on his own feet. The debate had caused a discord within the family; Lee Man Tat was struggling with a choice of difficulties given the conflicting views amongst his sons.

Having said that, Lee Man Tat understood well the urgency and peril of the case. He could not allow such an agitated atmosphere to persist, either in the family or the company. He therefore decided to stabilise the situation by first cutting the costs of operation.

Lee Man Tat's first step was to boost the morale of the staff. On 23 December 1999, he invited the senior management team of Nanfang Lee Kum Kee to his home in Repulse Bay, Hong Kong. Sammy Lee, Harry Yeung and Stephen Chung(4) were present amongst others. Lee Man Tat encouraged the team and asked them to persevere in the business:

> *If any of you bears no confidence to continue in the business, you can get up and leave now... I have absolute confidence in the company's future, and I will give you my full support. We will definitely overcome difficulties and get to a better development.*

As expected, no one had the guts to quit the struggle at such a critical time. They pledged to work together and try their best to confront the prevalent problems, and eventually they did bring the company back from the brink of collapse.

Lee Man Tat's next step was to send his trusted confidant, Tang Fook Chuen, to Guangzhou. Tang Fook Chuen's mission was to help streamlining Nanfang Lee Kum Kee's administration, as well as imposing strict budgetary measures to curb expenditure. Rationales behind these measures were first of all to safeguard its survival; secondly, to conserve limited resources as much as possible for its future take-off. The pragmatic and reliable "Chinese Superman" as described by Sammy Lee worked around the clock to heal the company from its wounds.

Tang Fook Chuen was particularly harsh on the company's spending over that extraordinary period. On one occasion he firmly rejected employees' request for the replacement of computers until they proved unusable. Salaries of staff had been frozen consecutively for   three years since 1998, not to mention awarding bonus. Under retrenchment measures, by 2000, the company had to cut back some of its staff, only to have them returned after the crisis. Such strict budgetary control did help the company to survive. Nanfang Lee Kum Kee endured the hardship for three years, and painful as it was, the company gradually regained its vitality.

Lee Man Tat's third step, in this regard, was to ask David Lee to approach HSBC in private. They were seeking an evaluation from the bank in case of a need to sell Nanfang Lee Kum Kee in the worst-case scenario. Lee Man Tat realised that even though health products industry remained very promising, it was nevertheless a rather high-tech and professional industry. It might therefore be fatal to the company in losing technological support of the First Military Medical University, while facing the ban by the state on direct selling. In the overall interests of the Group, Lee Man Tat must take a helicopter view to prepare several contingency plans in his attempt to get out of the crisis.

However, what Lee Man Tat worried most was not the survival of Nanfang Lee Kum Kee, but the intention of his youngest son, Sammy Lee, to leave Lee Kum Kee. Sammy Lee is in many ways a "Lee Man Tat Junior," in the sense that both are full of entrepreneurial spirit. The family's decision to wind up his successful new business of Kin Yat Kitchens still rankled in his mind. Similarly, he spent such a great deal of time and energy to run Nanfang Lee Kum Kee, and the business had proven its value and potential. He was thus most unwilling to see the end of it simply because it was struck by a sudden state policy change. Sammy Lee once proposed to give up all his shares of the parent company, i.e. Lee Kum Kee Group, in exchange for shares of Nanfang Lee Kum Kee. In other words, he wanted to quit the Group. It touched the nerve of Lee Man Tat who feared to face yet again a family separation in his life.

As the streamlining of Nanfang Lee Kum Kee's administration was in train, Sammy Lee dwelt on reflections over his life, career, and the family's destiny. Pondering on his father's indoctrination, as he recalled, had guided him to a deeper understanding of the philosophy behind "Si Li Ji Ren" and "putting oneself in others' shoes." When he further reviewed the development course of his family thus far and the way forward, he also had a new understanding of the meaning of a "family." An epiphany came to him in the relationship between himself and the family. Simply put, he came to accept that "We" were bigger than "I", and he must therefore abide by the long-term perspective of the family first. Moreover, a management philosophy connecting elements of "helicopter mindset", "putting oneself in others' shoes" and "Dao, Tian, Di, Jiang, Fa"(5) began to take a clearer shape in his mind during this period of thorough deliberation. It integrated the traditional and the modern management wisdom from both the Chinese and the Western perspectives, and gave him a greater sense of mission and motivation to lead the company out of troubles.

Sammy Lee's internal mind struggling ended at that moment of epiphany. He was no longer obsessed with the idea of breaking up from Lee Kum Kee Group, and single-mindedly devoted to resolve the company's crisis. Adopting a new approach of strategic thinking, Sammy Lee and his senior management team finally came out with their first Five-Year Plan in 2001, and subsequently led the company out of the trough to head for a swift revival.

Nanfang Lee Kum Kee was finally back on track after a couple of years of hard work. Tang Fook Chuen returned to Lee Kum Kee Headquarters in Hong Kong on completion of his mission, while the reborn Sammy Lee developed an even more comprehensive management philosophy of "Autopilot Leadership Model" and led the company to new heights. It was again a case of turning crisis into opportunity by the Lee family. In 2005, the state finally promulgated *Administration of Direct Selling Regulations* to regulate direct selling in Mainland China. It cleared the cloud of suspicion over the industry, and therefore allowed Nanfang Lee Kum Kee to take a big stride.

Lee Man Tat proved his great leadership once again in the face of the multi-layer crises in 1998: the threatened financial stability of Lee Kum Kee Group; the seriously crippled Nanfang Lee Kum Kee; and the possibility of a third family separation. These were all thorny issues and risks to Lee Kum Kee and the Lee family. Remarkably, Lee Man Tat coped with these challenges calmly and steadily with the realisation, at the same time, by Sammy Lee that "blood is thicker than water".

In fact, the crises of 1998 strengthened the relationship of family members after a conflict of opinions. Each of them learnt to appreciate better the differences between parents and children as well as amongst the siblings in their ways of thinking and matters of concern. It could be

Planning is crucial to the development of any country and enterprise. The constant and high-speed advancement of Infinitus is the result of its successive five-year plans, which were formulated through concerted efforts of its management and staff.

said that "Si Li Ji Ren," the motto long cherished by Lee Man Tat, has a new connotation after the crises of 1998.

The post-1998 era not only saw another round of growth of Lee Kum Kee Group, but an amazing surge of Nanfang Lee Kum Kee's business comprehensively. By 2009, in order to better develop the market of health products, the management team decided to rename Nanfang Lee Kum Kee as "Infinitus (China) Company Ltd.", so that two distinctive brands of health products and sauce products could be effectively differentiated.

Infinitus subsequently adopted a new brand and image, followed by a heavy investment to further promote its business. Since then, it has rapidly risen to be one of the leading direct sales companies in Mainland China. Infinitus sets its mission as "advocating the premium Chinese health regimen; and nurturing healthier lives with balance, affluence and harmony." In pursuance of this mission, it aims to promote Chinese herbal medicine and Chinese health regimen to meet the demands of a new era.

Today, with a new name, new trademark and new goal, Infinitus on the one hand trains its sales teams to be more professional, while on the other hand increases its property investment and has acquired landmark buildings in Hong Kong, Guangzhou and Shanghai. These buildings have since become its most powerful representation of image and display of strength. Infinitus's annual turnover today has exceeded HK$ 20 billion. In fact, both the Sauce Group and the Health Products Group of Lee Kum Kee have achieved milestone developments since the new millennium. Their accomplishments surely please Lee Man Tat; but the entrepreneurial minded octogenarian has not slowed down his pace and continues to charge ahead to lead Lee Kum Kee to the world.

The initial logo of Nanfang Lee Kum Kee (above) and the new logo of Infinitus (below). The rebranding of Nanfang Lee Kum Kee as Infinitus was a milestone in the company's development, with a logo denoting "infinite possibilities from collective strengths."

Go Worldwide

As mentioned earlier, from its inception in 1888 to the 1990s, Lee Kum Kee relied heavily on the overseas Chinese markets. One of the major factors contributing to its popularity abroad was the "cultural taste" of the overseas Chinese who never stopped yearning for Chinese cuisines even in faraway lands. Lee Kum Kee, under the three generations of the Lee family, represented by Lee Kum Sheung, Lee Shiu Nan and Lee Man Tat, has consistently embraced its high-quality approach to pursue the overseas Chinese markets, thus achieving outstanding reputation of quality. Lee Kum Kee's oyster sauce is particularly notable and visible amongst the overseas Chinese. It has become a key sauce for many Chinese dishes, as well as an indispensable condiment on dining table of many Chinese families.

The prevalence of Lee Kum Kee's oyster sauce in the overseas Chinese communities is obvious. It could be found in different types of grocery stores and supermarkets worldwide, and this is an amazing phenomenon frequently reported by a variety of cultural, food and tourism magazines. This was also noted by He Xiansong, former Mayor and former Party Secretary of Xinhui City, on his first official tour abroad in 1982, which had in fact featured Lee Kum Kee's outstanding achievement in overseas territories:

> I went abroad for the first time in 1982. In a little town of Switzerland called Chur, I found merely two Chinese products in a local grocery shop: one was Lee Kum Kee's oyster sauce from Hong Kong, and the other were table-tennis balls from Mainland China. I was shocked. From then on, I vowed to usher in Lee Kum Kee's investment to Xinhui.

Meng Suhe, President of the Chinese Institute of Food Science and Technology (CIFST), mentioned a similar experience and feeling in her interview with us, which also revealed that Lee Kum Kee did market

Lee Man Tat and his four sons on visits to come to grips with the market penetration of Lee Kum Kee in the American supermarkets in the late 1980s.

itself around the world:

> *Upon the invitation of the French Embassy in 1990, I attended the "Paris International Agricultural Show" (Salon International de l'Agriculture) where Chinese products were rarely seen. I finally found the booths of China National Cereals, Oils and Foodstuffs Corporation at a corner, and they handed out spicy sesame oil to visitors. But its sesame oil had dripped through the bottle on my hand even before I got to the exit. As a Chinese, I felt particularly sad about that... But later, I saw people who came out from an exhibition hall each holding a very beautiful bag, with "Lee Kum Kee" printed on bags. These were Chinese characters, which I felt so familiar, thus I walked towards the hall. I then saw Lee Kum Kee's typical and well decorated Chinese-style archway, with its exquisite products on display. Its booth was patronised by many foreigners. I was still young then, with very limited experience of going abroad, and the scene immediately prompted a sense of pride in me. Also, when I was in the United States, the most delicious Chili Garlic Sauce of Lee Kum Kee was available in every Chinese restaurant there. And so naturally, Lee Kum Kee left a good impression in my mind.*

Meng Suhe's personal experience and observation in the 1990s served as an excellent testimony of Lee Kum Kee's reputation and its lion's share in the overseas Chinese market. It also explained why Lee Man Tat, a man who always projects constant entrepreneurship, would turn his eyes squarely on China for further advancement of Lee Kum Kee. And given the integration of China in a globalised world, Lee Man Tat's next step from China would naturally be the international market. Such an international market would go beyond the overseas Chinese communities in which Lee Kum Kee has dominated their sauce markets for nearly a century.

Lee Kum Kee endured severe hardships during the crises of 1998.

When they were over after a series of re-structuring and hard work, Lee Kum Kee has then entered another stage of high-speed growth since 2000. The advancement is mainly due to its substantial expansion in the Mainland market, instead of the North America's.(6) The fact is, Lee Kum Kee has long dominated the Chinese-style food service markets in North America, Japan and South Korea.

Lee Kum Kee has also performed considerably well in the European market since the 1990s, particularly in the UK. It also made good progress in the Oceanian countries of New Zealand and Australia. It nevertheless encounters strong challenges from the entrenched local sauce brands in the Chinese markets of Southeast Asia, and is therefore unable to make any breakthrough yet. In short, today, Lee Kum Kee's products of over 200 choices have become popular in more than a hundred countries and regions across the five continents.

It is also worth noting that Infinitus has become the Group's new engine of growth, with a soaring sale of its health products. Lee Man Tat had in the early days set a tough goal for Lee Kum Kee to achieve "where there are Chinese people, there is Lee Kum Kee". It has long realised this and achieved a worldwide popularity of "where there are people there is Lee Kum Kee". And, beyond all dreams, it has now even achieved the eventuality of "there is Lee Kum Kee, even where there are no people." A number of landmark developments could serve as footnotes to the outstanding progress of Lee Kum Kee:

· In 2008, Lee Kum Kee was appointed official supplier of sauces for the Beijing Olympics;

· In 2008, Lee Kum Kee reaped the "International Union of Food Science and Technology – Food Service Award" jointly presented by the International Union of Science and Technology and the Chinese Institute of Food Science and Technology;

· In 2010, Lee Kum Kee was appointed as "Official Qualified Suppliers of Raw and Auxiliary Materials in Catering Category" at the Shanghai World Expo;

· In 2012, Lee Kum Kee was selected as "Partner of China Space Industry" to provide sauces for "Shenzhou IX" manned spacecraft projects;

· In 2013, Lee Kum Kee was again selected to provide sauces for "Shenzhou X" manned spacecraft projects;

· In 2015, a Lee Kum Kee's product was scrutinised and selected by the United States space agency NASA for the consumption of its astronauts;

· In 2016, Lee Kum Kee was named "Classic Century Food Brand" by the Hong Kong Food Council;

· In 2016, Lee Kum Kee was selected for the third time as sauce provider for "Shenzhou XI" manned spacecraft projects...

Among the achievements listed above, Lee Man Tat is drastically proud of the fact that Lee Kum Kee had repeatedly appeared in outer space, showing that "Lee Kum Kee could even be found in no man's space." The story began in 2012 when Lee Kum Kee was first selected as "Partner of China Space Industry" to provide five types of sauces for the astronauts of "Shenzhou IX" space mission. It was subsequently selected to serve the astronauts of "Shenzhou X" (2013) and "Shenzhou XI" (2016) as well with six types of sauces.

Space food production is scrutinised against drastically more stringent standards than food served to general consumers. It accentuates absolute food safety, longer shelf-life, as well as compatibility with varying space conditions and the individual needs of the astronauts.

The repeated appearance of Lee Kum Kee's products in outer space following the three voyages of the Shenzhou IX, X, XI, as well as in the International Space Station, was indeed unprecedented.

Lee Kum Kee's status as a space food supplier proved that its quality control and standards have reached the industry's highest level. Lee Kum Kee paid meticulous intention to the manufacturing of sauces tailor-made for the space missions of Shenzhou, and had accordingly selected Guizhou Black Bean Chilli Sauce, Sichuan Hot & Spicy Sauce, Peanut Flavored Sauce, Thai Sweet Chilli Sauce, Chilli Bean Sauce and Sweet Sauce for the astronauts. This collection covers a diverse taste profile from sweet, sour, salty, to hot and spicy, delighting their palates to reward them for their arduous space assignment.

There is one anecdote to note in this regard. In November 2015, while on his one-year mission in the International Space Station, American astronaut Scott Kelly prepared a Thanksgiving feast for his Russian and Japanese partners with his favourite sauce – Lee Kum Kee Sriracha Mayo. It was only after this event was widely shared on the social media that Lee Kum Kee realised that one of its products had been scrutinised and selected by the United States space agency NASA for consumption by its astronauts in the outer space.

Over the past 130 years, Lee Kum Kee set off from a secluded corner in South China – Nanshui, and traversed across the Pearl River Delta, followed the steps of overseas Chinese to North America, Australia, Japan, and Southeast Asia. Finally, the legend returned to where it originated from, China, and broadened its stride to go worldwide. Apart from the Tai Po Headquarters in Hong Kong, the two groups of Lee Kum Kee have established production bases in the United States (Los Angeles), Malaysia (Kuala Lumpur), Guangdong Province (Xinhui, Huangpu) and Liaoning Province (Yingkou), forming a pretty globalised network of production. The following examples testify the Group's accelerated pace of internationalisation as well as its scale of advancement in recent years:

· In 2009, it acquired "Infinitus Centre" in Guangzhou;

· In 2010, it acquired Vicwood Plaza in Central Hong Kong and renamed it "Infinitus Plaza";

· In 2014, its large production base in Yingkou, Liaoning Province commenced production;

· In 2015, it acquired Shanghai Corporate Avenue 3 and officially named it "Infinitus Tower";

· In 2016, it started the construction of "Guangzhou Infinitus Plaza" – the new headquarters of Infinitus (China) Company Ltd. in Baiyun District, Guangzhou;(7)

· In 2016, it acquired an office building in Central, Hong Kong and renamed it "Lee Kum Kee Central";

· In 2016, it acquired an office building in Xuhui District, Shanghai and named it "Lee Kum Kee Building";

· In 2016, it acquired the office building of 3 Harbour Exchange in London;(8)

· In 2017, it acquired the landmark commercial building at No. 20 Fenchurch Street, London;

· In 2018, it acquired a plot in Guangzhou South Railway Station Commercial Hub to build Lee Kum Kee Sauce Group's South China Sales Headquarters...

Lee Man Tat's leadership is indisputably the key to Lee Kum Kee Group's persistent advancement through all sorts of hardships it overcame in almost half a century as well as to its prominent achievements. Lee Man Tat is also highly respected and praised throughout the industry, worldwide, and by the public, for his continuous and tremendous contributions to the society, which was indicated in innumerable international awards he had received over the years. Some of the following titles and compliments are best explained:

· In 2004, he was awarded "Hong Kong Business Mastermind" by the Hong Kong General Chamber of Commerce;

· In 2006, he was appointed "Advisory Professor" by Tsinghua University;

· In 2008, he was awarded "The Most Caring Charity Individual" by the Ministry of Civil Affairs;

· In 2008, he was inducted into Babson College "Academy of Distinguished Entrepreneurs" in the United States;

· In 2009, he was awarded "Industrialist of the Year" by the Federation of Hong Kong Industries;

· In 2010, he was granted the "Bronze Bauhinia Star" by the Government of the Hong Kong Special Administrative Region;

· In 2012, he was awarded the "Southern - Top Ten Chinese Philanthropists" by the Overseas Chinese Affairs Office of the People's Government of Guangdong Province;

· In 2013, he was elected "Leader of the Year 2012" in the "Commerce and Industry / Finance Category" organised by Sing Tao News Corporation;

Lee Man Tat was awarded the "Bronze Bauhinia Star" by the Government of the Hong Kong Special Administrative Region in 2010, for his extraordinary achievements and tremendous contribution to Hong Kong.

· In 2016, he was honoured the "Outstanding Achievement Award" by the Hong Kong Food Council;

· In 2017, he was elected as the "Global Top Ten Outstanding Cantonese" at the Third Global Conference of the Cantonese;

· In 2018, he was granted the "Silver Bauhinia Star" by the Government of the Hong Kong Special Administrative Region;

· In 2018, he was granted the Lifetime Achievement Award in the Hong Kong Business Awards 2018 (organized by DHL and the South China Morning Post) for his sustained business and social contributions to Hong Kong...

Albeit showered with myriad awards and compliments, Lee Man Tat is far from complacent. The octogenarian remains active and keeps deliberating over strategies that he could adopt to further propel Lee Kum Kee's advancement. He is also sincere and amiable in his interactions with others, and such humility has earned him heartfelt respect. Sima Qian once praised Confucius in his *Records of the Grand Historian (Shi Ji)* as follows: "The poem says: 'Look up at the high mountains and walk along the wide road.' Although I cannot reach this level, my heart desires to be there." People who have experience to interact closely with Lee Man Tat may similarly share the same feeling of respecting the grand old man as to look up to high mountains.

Conclusion

There is a strong consensus amongst the closest friends of Lee Man Tat that he is a man who is incessantly making breakthroughs in his life, and therefore reaching new heights one after another. Lee Man Tat's immense dynamism since his adolescence has advanced Lee Kum Kee from a small-size oyster sauce store equipped with family workshop to an eminent brand well-known to the Chinese worldwide. The company's sales networks comprehensively cover overseas Chinese communities in North America, Australia, and Southeast Asia. Lee Man Tat's aggressive entrepreneurship is by no means subdued in his elderly days, as he keeps thinking about starting new businesses and exploring new fields.

As Mencius said: "The superior man makes his advances with earnest learning and by the proper course, wishing to get hold of it in himself. Having got hold of it in himself... he could then tap into these endless resources from within himself." Lee Man Tat's insights and his timely grasp of business opportunities were based squarely on his rich and legendary life experiences as well as his powerful and resourceful networks. They allow him to perform amazing feats one after another. He led Lee Kum Kee in breaking through the vast Chinese market, subsequently to the world and even to the outer space. Lee Kum Kee's space voyages remain a popular topic in the media today.

For all his stature, Lee Man Tat treats people modestly and generously. He is amiable and admirably unassuming, frank and open-minded. Such personal attributes have won him not only awards, but heartfelt praises and respects across his family, his friends, the business world, and ultimately the community at large.

1   Tony Mok joined Lee Kum Kee in 1992 as a management trainee. He has been involved in the production of soy sauce ever since.

2   Fibreglass is a common type of fibre-reinforced plastic (FRP) using glass fiber. The advantages of fibreglass include lightweight, corrosion-resistant, water-proof, good insulation and durable, thus it could be an alternative to steel in manufacturing machine parts as well as automobile and ship shells.

3   Lam Yu is currently Senior Vice President, Lee Kum Kee Health Products Group; Global CEO of Infinitus; Vice Chairman cum Managing Director of Infinitus (China) Company Ltd.

4   Stephen Chung is currently Senior Vice President of Lee Kum Kee Health Products Group.

5   Dao refers to a common goal and direction. Tian means the external variables. Di refers to the internal variables. Jiang means choosing the right talent, and Fa refers to method (Lee, 2016: 91-92).

6   Today, the Chinese market constitutes around 60% of the turnover of Lee Kum Kee Sauce Group.

7   Guangzhou Infinitus Plaza is one of the splendid artworks by the world-renowned architect Zaha Hadid, also known as "Queen of the Curve". It has a great sense of beauty with "infinite rings" design from the inspiration of "∞".

8   3 Harbour Exchange is located next to Canary Wharf on the bank of River Thames in London. The 11-storey property covers a total floor area of 90,000 square feet, with one floor to be used as the Europe regional office of Lee Kum Kee Sauce Group.

Lee Man Tat shared the key of Lee Kum Kee's success with the audience in the lecture hall of Tsinghua University in Beijing in 2006.

PREEMPTING FAMILY CRISIS

Now the fourth generation is successfully running the family business together. In the future, members will relinquish the work to the fifth generation. With the number of people getting bigger, relationships will become more complicated, opinions more diverse and communication more difficult. This is a challenge to the family. But I firmly believe that if we embrace the family core value of Si Li Ji Ren, all problems can be resolved because unity is strength.

Lee Man Tat

INTRODUCTION

Every life has its cycle of birth, growth, maturity, aging, and death. It applies to the development of any family and any business as well. In the eyes of the Chinese, a big family with "hundreds of children and thousands of grandchildren" is certainly a blessing; while on the other hand, they also recognise the tendency of division once a family is getting bigger, regarding the process as natural as the branching out of big trees or "the full-fledged birds leaving their nests." However, the "rule of division" sensibly acknowledged by a patriarch may not be his preferred outcome emotionally. Many Chinese parents still hope that their children and grandchildren could all be around harmoniously, and that there are affection and respect among siblings. They may also worry that a family separation would hurt the feelings of one another and weaken its strength, and will therefore be unfavourable to its prosperity. Likewise, the first and the second family separations of Lee Kum Kee in the 1970s and 1980s were reflections of the natural "rule of division," yet they saddened Lee Man Tat very much as he always cherishes the relationship with family members and strives for a harmonious family.

The strong emotional impacts of the two episodes of family separation on Lee Man Tat had prompted him an intellectual deliberation in his autumn years. Indeed, his business kingdom was getting ever larger and his children had performed remarkably well in their respective positions, but the 1998 crisis of Nanfang Lee Kum Kee triggered by drastic external policy changes revealed the diverse opinions and goals as well as the different pace of advancement within the family, thus the inclination towards discords and even conflicts. When Sammy Lee proposed the establishment of a Family Council which was then agreed by his siblings, the astute Lee Man Tat immediately realised that if relations and emotions within the family failed to be handled properly, it might trigger the third family separation. He therefore engaged in attempts to nip the problem in the bud to preempt such family crisis, and subsequently came out with a brand new approach to safeguard family inheritance.

Strengths and Weaknesses of Family Business

Chinese family businesses have long been negatively labelled as small-size enterprises with backward management. In both the Chinese and the Western societies, people generally share a perception that family businesses are conservative and rather disorderly in their organisation, and therefore difficult to develop themselves bigger and stronger, not to mention transform themselves into modern institutions to achieve world-class competitiveness. Albeit they are highly dynamic, efficient and resilient, their negative image is so deep-rooted that people still regard them as plagued by troubles rather than blessed with merits.

As a man dedicated his life to Lee Kum Kee, Lee Man Tat surely understands both the strengths and weaknesses of a family business. The family and its business, if intertwined as an entity of common destiny, could release immense power to advance the business as its success could bring glory to the ancestors and benefit the descendants. Such a motivation to excel and incessantly pursue breakthrough is unmatched by any governmental or public institution since bureaucrats concerned tend to work as a matter of course according to institutional rules and regulations with less emphasis on relationship management.

The close relationship between the family and its business implies that they will support one another in the face of difficulties as they share a common destiny. More importantly, because the business and its brand represent the best interest and glory of the family, and that the family would strive its very best to sustain the brand for generations to honour the ancestors, it will then consider issues from a long-term perspective of development. Family businesses therefore tend to pay more attention to the quality of their products and their relationship with employees and business partners, as well as their social responsibility. These are positive attributes again unmatched by non-family businesses.

Family business nevertheless has its intrinsic drawbacks in terms of the perfect integration of the family and its business. Amongst others, the family would place its interests and goals above its business, and emotional relations amongst the family members would inevitably extend to its business. These could affect the management and development — in particular financial and personnel arrangements — of the family business, leaving observers a negative perception of nepotism and over-emphasis on interpersonal relations that highlight differences between family members and non-family members. In short, to the general public, family business is simply too family centred, and such nature hinders its course towards standardisation and openness.

Internal conflicts and cleavages within the family are always more sensational to the public and deeply worrying to families concerned. Amongst others, the issue of succession after younger generations have joined the business is the most intractable. A variety of problems would become more complicated when different life cycles overlapped and intertwined, and the piling up of resentments could easily trigger disputes and conflicts. These conflicts, if tackled inappropriately or fuelled by miscommunication, may cause serious power struggles and division, and therefore inflict unhealed wounds on both the family and its business.

There is no doubt that Lee Man Tat hates to see the repeat of a family separation. The two previous cases of family separation began similarly with internal contradictions and conflicts, and burst out with the accumulation of mutual resentments later, eventually leading to a split and an estranged relationship among family members. These experiences were so painful and unforgettable to Lee Man Tat that he vowed to avoid the repeat of such tragedy to his descendants.

As mentioned in Chapter 8, the Lee family, in the face of a series of crises in 1998, had experienced internal conflicts due to diverse opinions of its members in tackling challenges, which, at one point, even prompted Sammy Lee to consider an opt-out from the family business to stand on his own feet. It was an alarm of yet another family separation and aroused a deep fear in Lee Man Tat, prompting him to take actions on "preventive treatment of disease" to ensure the sustainability of Lee Kum Kee. The challenge was not only to further advance the business, but to handle its succession as well, so that unity of the family could be strengthened.

However, this is by no means an easy task to deal with. There is a popular saying in the Chinese society that "wealth never survives three generations." It reflects the inductive wisdom of the common people that first and foremost, a rich generation tends to relax and slow down their pace in creating wealth; and secondly, the principle of "equal inheritance" embedded in the Chinese culture has made family separation a norm rather than an anomaly.

In fact, in order to break the curse of "wealth never survives three generations" and sustain a harmonious big family for generations, numerous Chinese families over thousands of years had tried their best to overcome the cultural and institutional hurdles concerned. There was first of all, an effort to emphasise and instill values of filial piety, brotherhood, loyalty and trustworthiness that could facilitate family unity and harmony in future generations. The second effort was to set up key family precepts to restrain behaviours of their descendants. The third effort was to stipulate cohabitation and estate sharing amongst family members, with the rule that family possessions should not be divided or sold to non-family members. The fourth effort was to establish charity Yizhuang (義莊), which could both support extended family members and benefit local communities, therefore ensuring a good karma to the descendants. These efforts and others, however, seldom successfully help families to overcome difficulties and challenges in passing down their businesses and wealth through future generations. Lee Man Tat was undoubtedly well aware of such harsh and undeniable reality for long.

That said, Lee Man Tat showed his consistent daring and resolution once again in this regard. He astutely spotted the problem in its infancy and tried to nip it in the bud, searching for "preventive treatment of disease" in his lifetime. With his tenacious character, he was not put off by difficulties and was determined to find solutions to achieve what might appear impossible to others. And in the search of a solution, Lee Man Tat was pragmatic, unassuming and able to humble himself to achieve anything that works, regardless of ideology. In that sense, he was in line with Deng Xiaoping's pragmatism of "regardless of white cats or black cats, it is a good cat if it can catch mice." In short, Lee Man Tat fought to overcome fate with perseverance once he became aware of a problem.

Significant Attempt in Creating Family Council

When the divisive forces were spotted within the family, Lee Man Tat was already in his autumn age. As a successful and seasoned entrepreneur with profound life experience, he had long enjoyed an esteemed social status, and his values and behavioural patterns were already deep-rooted. It could be imagined that for other patriarchs placed in similar situation, they might simply respond with models deemed effective in the past, rather than incorporating new elements for a breakthrough. But Lee Man Tat, the then septuagenarian, was determined to solve this grave issue which might shake the foundation of Lee Kum Kee. He took immediate actions to explore the issue together with his children, free from personal pride and prejudice, as they tried out different solutions. What was more crucial was his openness to incorporate new elements in creating the "Family Council," the pioneering institution amongst Chinese family businesses.

Lee Man Tat mentioned in interview that his constant way of thinking was to place family in a far higher position than the enterprise, and that the key to the continuation and expansion of any family business was family harmony, unity, and right moral values. It is in fact a very traditional concept of "when brothers unite, their strength is powerful enough to break metal." Thus, after experiencing the highly tortuous family separations in 1972 and 1986, as well as the intention of Sammy Lee to leave Lee Kum Kee over 1998 to 2000, Lee Man Tat began to pay attention to the succession and sustainability of the family and its business.

In the era of globalisation, companies must strive to become bigger and stronger for better survival instead of struggling alone. The fourth generation of the Lee family also realised such a trend and understood the importance of family harmony in further advancing their family business and ensuring its sustainability for future generations. Initiated by Sammy Lee, the fourth generation began to collectively participate in seminars on family business organised by globally renowned universities such as Harvard and Tsinghua since 2000, contemplating and searching for a power structure that could facilitate long-term stability and continuity of family business through international case studies and academic theories. Sammy Lee shared and discussed such ideas and their objectives with Lee Man Tat, and soon got his positive response and endorsement. They then proceeded with more thoughtful attempts and learning, and decided to set up Family Council when they had 60% to 70% confidence of success (the "6677 Principle").

Sammy Lee further explained that the Family Council was collectively explored, deliberated and formed by the fourth generation with blessings from Lee Man Tat and Choi May Ling. It was a process similar to their previous decision to learn how to make soy sauce for the market. They likewise encountered various problems, and it was only after intensive discussions and practices that they finally reached that unanimous decision.

One important issue that they repeatedly touched upon was the solution for internal conflicts. They acknowledged the merits of family business in terms of its strong emotional bonding amongst members, but that could be a demerit as well. When the number of family members increased, relationships became more complicated, affectional appeals might therefore be no longer enough to solve problems and meet the challenges. A modern institution was needed to better deal with differences and conflicts, so as to keep the business stronger and enable it to compete with other world-class enterprises. In other words, it was necessary to come up with a new approach or arrangement with positive energy to solve internal conflicts, rather than simply resort to the old-fashioned moral or emotional appeals in a passive way.

Family Council as an institution has in fact been adopted by certain prestigious families in the West. However, it has received much less attention in the Chinese society, as many elements of such an institution had long been practised under different names throughout thousand years of Chinese history. In other words, these are nothing new to the Chinese. For instance, the "Family Constitution" is no different from the ancient family precepts; and for those trust funds created by many families with the main purposes of charity, education and safety net for the family members, they are by and large similar to the traditional Chinese institution of Yizhuang, which were charities established by prominent families between the Song and the Qing Dynasties to support their clansmen and local communities, offering charitable services such as orphanages, free schools, disaster relief, and coffin homes.

Nevertheless, the institution of Family Council advocated by the West does have its appeal and significance:

The emerging capable members of the fifth generation have joined the "Family Council" one after another. From left along the conference table: David Lee, Elizabeth Mok, Charlie Lee, Jason Lee; standing in front of the table: Eddy Lee, Brian Lee; from right along the table: Sammy Lee, Professor John Ward, Charles Mok, Andrea Lee.

· First, rules and regulations of the Family Council are not defined by the patriarch alone in a top-down manner, but collectively shaped through bottom-up discussions participated by both the patriarch and the younger generations;

· Secondly, its rules and regulations are written in the company's Articles of Association or reflected in details of the Trust, and are therefore legally binding. It makes them distinctively different from family precepts.

· Thirdly, Family Council as a "modern institution" is more acceptable to the younger and overseas-educated generation in the line for succession, thus encourages their participation;

· Fourthly, Family Council is an institution more systematically organised. While it is led by Council Members, there are specialised units within the framework which carry different functions. Moreover, the institutional design of Family Assembly and Family Constitution serve as mechanisms of check-and-balance to ensure accountability; whereas the "one person, one vote" rule safeguards the key element of democracy in the system, instead of dominated by the patriarch under the traditional paternalism.

In a nutshell, Family Council is regarded as a modern institution which broadens the participation of family members in succession arrangement, compared to the traditional way of counting on the will of the patriarch. It could therefore allow better and easier interactions amongst family members from different generations.

It is precisely such attributes of the Family Council which appeal to younger generations and prompt their participation. Lee Man Tat, the open-minded and informed patriarch, was willing to proceed with the pioneering attempt of founding the Family Council of Lee Kum

Kee in 2002, which was in fact the first amongst the Chinese family businesses.

Sammy Lee recalled that when they finally reached the consensus to establish the Family Council and were preparing for the first meeting, they decided to try something special, which was to hold the meeting at the private yacht of Lee Man Tat instead of in the office. Moreover, they did not set an agenda for the meeting in advance, but encouraged members to walk around freely on the yacht for a literally free discussion, so that everybody could really speak his/her mind for optimal solutions to issues concerned.

Charlie Lee frankly admitted that at that time, "we didn't really know what we could do yet," as they were sailing in uncharted waters. Sammy Lee added that thereafter the Family Council met regularly every three months, each for a period of four days. The first three days of the meeting are restricted to core members of the Council only, while the last day is open to all members of the family. In sum, it is a channel for members to freely express their views for a brighter future of the family, and the full participation of all members will facilitate internal communication and improve relationship.

David Lee pointed out that Family Council, as the highest decision-making body of both the family and the enterprise, highlights collective leadership, and that obviously differs from Lee Man Tat's strong leadership over the years. Moreover, rules set by the Family Council will become Family Constitution, which governs behaviours of all family members, including Lee Man Tat and Choi May Ling. As for the number of Council members, it has since expanded from only seven — Lee Man Tat, Choi May Ling and their five children — to the present number of nine. Lee Man Tat and Choi May Ling had voluntarily withdrawn from the Council on 1 January 2017, and were subsequently appointed "Honorary Founding Members." They vacated places for members from the fifth generation: Brian Lee, Charles Mok, Jason Lee and Andrea Lee. The main purpose of such reshuffle was to get the fifth generation familiar with the running and decision-making of the family and its business as soon as possible, so as to equip them for succession in the future.

Lee Man Tat is happy to see his five children perform outstandingly in their respective positions yet remain low profile; and he is very proud of the fact that there is no more domestic fight and conflict within the Lee family since 1986, which is in fact a rare record compared to other prominent families in Hong Kong. Indeed, Family Council as an idea was not initiated by Lee Man Tat, but he had obviously pondered on problems embedded in traditional succession arrangements before he came to the conclusion that this was rational and feasible, and could strengthen internal communication, eliminate misunderstandings as well as enhance cohesion within the family.

We had learnt from interviews that the main consideration of Lee Man Tat was to gradually devolve power as well as responsibilities to his children through the Family Council. He also intended to embed several core values that he cared most about into the Family Constitution. For example, the Family Constitution should expressly stipulate that members should not divorce and should avoid involving in any extra-marital affair, lest they be expelled from the Family Council and deprived of all duties. According to Charlie Lee, Lee Man Tat initially also stressed that family members should get married early (NO late marriage). But the proposal was not supported by other members of the Family Council, and hence failed to be incorporated into the Family Constitution.

Other articles of the Family Constitution which also have profound implications include the rule that in future, members who want to join the business are required to have working experience in other

Lee Man Tat and his five children were freely sharing ideas during
their discussion at a meeting of the Family Council.

companies for 3 to 5 years. Also, they must, like other job seekers, pass the assessment and start at the grassroots level. If their work performances are not up to the standard, they will be given a second chance; but if they fail to improve, they should be dismissed like others. Another significant point to note is that Lee Kum Kee Group adheres to family holding, and hence the Chairman must be from the family, and if someone has to quit the Group for any reason, his or her shares must be bought back by the Group. The Family Constitution also has a retirement rule that stipulates compulsory retirement from the Board of Directors and managerial positions at the age of 65, and from the Family Council at the age of 70.

The division of labour in the Family Council is rather painstaking. There are five units under the Family Council, namely Business (i.e. LKK Sauce Group and LKK Health Products Group), Family Investment, Family Foundation, Family Learning & Development Centre, and Family Office, each has its own responsibility. Regardless of the details, even an outsider can tell from such a governance structure where business is merely one part of the family, and the notion of "we" bigger than "I" is reflected in its distinct sense of collectivism. In Lee Man Tat's own words, the Family Council and the Family Constitution were adopted to "tie" family members together.

Lee Man Tat is eager to see this creative institution of succession in the history of Chinese family business to move forward side-by-side with the Lee Kum Kee Group. The speech he delivered to the Family Council on 29 December 2012 deserves to be highlighted:

> I am willing to do anything to achieve family unity and harmony. In my experience, the family has to pay heavily for any family conflict. Because of differences in views on family business development, my brother stopped contacting me and this inflicts much pain on me. I do not want to see my descendants suffering from the same experience.

The Family Governance Structure of Lee Kum Kee

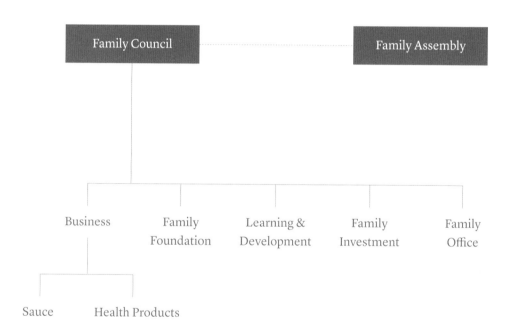

Therefore, I urge you wholeheartedly to put the family above everything else and, in the spirit of Si Li Ji Ren, to take a helicopter view to maintain the common interest of the family. Without a harmonious family, we cannot sustain the family or the family business.

Admittedly, the effectiveness of Family Council as a co-governance platform remains to be seen, but the fact is Lee Man Tat, despite his old age, still endeavours to preempt crisis, which is again a reflection of his ability to see the big scene from the small one. He is also praiseworthy in accepting new ideas with an open mind and trying them out. In fact, Family Council as a modern institution and platform does allow younger generations to openly discuss issues and eliminate misunderstandings and resentment in a timely manner, thus facilitating family harmony.

A Model of Family Business Succession

After serving Lee Kum Kee for half a century and experiencing two episodes of family separation, Lee Man Tat began to seriously think about a better mechanism for resolving internal conflict in the late 1990s. He realised that without it, the sustainability of the family and its business would be in jeopardy.

The creation of Family Council was the most important result of such deliberation. This institution, though ostensibly a modern design, bears various connotations of Chinese traditional culture, such as its emphasis on kinship, filial piety, brotherhood, loyalty, and trustworthiness as well as open discussion amongst individual families of the Lee family clan.

Exactly because Family Council is a modern institution with Chinese cultural connotations, it is acceptable even to patriarchs imbued with Chinese traditional culture, hoping that it could create a favourable environment and platform for future generations to better communicate and interact. On the other hand, in the eyes of those younger generations influenced by western education, Family Council may sound modern with an emphasis on both individual rights and responsibilities, and its equal voting system is particularly appealing to them. They may therefore actively participate in the institution and contribute more to the family business.

In the early stage of implementing Family Council, Council Members frequently argued on its rules and regulations. The previously mentioned "Three Forbidden Rules" (NO late marriage, NO extra-marital affair, and NO divorce), for instance, were not fully enacted due to lack of consensus amongst the Council Members on early marriage. The debate nevertheless revealed the democratic nature of Family Council which allows open discussion, rather than dominated by the patriarch. It also showed the breadth of mind of Lee Man Tat in devolving his power for joint governance with the younger generations.

After the early stage of exploration, the Family Council began to operate smoothly and has since become an important platform for planning, deliberation, communication, training and networking. Given such standardised platform and mechanism, family members can now enhance their understanding of the family and its business through interactive communication, and have consequently improved their mutual trust and solidarity. More importantly, negative emotions within the family could be managed earlier and better. The Family Council has greatly reduced the risk of succession and enhanced cohesion of the family. The Lee Kum Kee Group, led by Lee Man Tat, has created an important model of succession for Chinese family businesses.

Lee Kum Kee's approach to overcome intensive disputes and conflicts during the process of succession has drawn extensive attention from the Chinese society and beyond. The media are eager to invite Lee Kum Kee to share its successful experience. Eddy Lee once mentioned that even some of the prominent families in Hong Kong had called him for sharing and learning.

Lee Man Tat was invited by Tsinghua University to deliver a public lecture on 15 May 2007, in which he shared his wisdom on running business and family governance and expounded on his experience of succession. To this end, Lee Man Tat succinctly highlighted the importance of "Si Li Ji Ren", "think from others' perspectives" and "the spirit of constant entrepreneurship" (please refer to Chapter 10). He also stressed the significant power of family harmony and unity, and naturally touched on how the Family Council could promote family cohesion and eliminate differences and conflicts.

The more noteworthy passage came from Lee Man Tat's speech at the meeting of Family Council on 29 December 2012. Apart from highlighting "co-governance" as the core of its institutional design, he touched on the reality of a bigger family and the subsequent challenges of more diverse viewpoints. Lee Man Tat sharply pointed out that its successful implementation lies squarely on whether its members could hold the family's core value of Si Li Ji Ren:

> *Now the fourth generation is successfully running the family business together. In the future, members will relinquish the work to the fifth generation. With the number of people getting bigger, relationships will become more complicated, opinions more diverse and communication more difficult. This is a challenge to the family. But I firmly believe that if we embrace the family core value of Si Li Ji Ren, all problems can be resolved because unity is strength.*

Apart from Lee Man Tat, his five children — Eddy Lee, Elizabeth Mok, David Lee, Charlie Lee and Sammy Lee — also frequently receive

Lee Man Tat delivers a speech during a visit to Tsinghua University in Beijing.

invitations from prestigious universities and institutions to share the wisdom of Lee Kum Kee's development and succession. These include Harvard and Stanford Universities in the United States; University of Cambridge in Britain; the Mainland Chinese universities of Tsinghua, Peking, Zhejiang and Sun Yat-sen; as well as the Chinese University of Hong Kong and Hong Kong University of Science & Technology. Composition and operation of the Family Council, as well as its mechanism for resolving conflicts, were invariably the focuses of their sharing and discussions. It is also worth noting that in recent years, certain research institutes specialising in company governance have used Lee Kum Kee, in their comparative case studies with related enterprises in Europe and America. (Sanaa, 2008)

The Lee family members are willing to share openly their experiences with the academia, the business world and the general public, if time and other conditions allow. They hope that such an institution, introduced by their family to the Chinese society with its integrated elements of modernity and Chinese traditional culture, could be put to better practice, so that more Chinese family businesses can sustain their heritage from generation to generation, and that the curse of "wealth never survives three generations" would no longer prevail.

Conclusion

We had mentioned in Chapter 2 that when Lee Man Tat was born, his parents named him "Kin Tai," which literally means "watch (him) grow up big." Their pious hope was to have the blessing of the heaven for his healthy growth. Yet intriguingly, the name unexpectedly revealed the most outstanding ability of Lee Man Tat: seeing the big scene from the small one. Such ability has enabled him to achieve a highly extraordinary life experience and career. As he could often sense and grasp opportunities timely in crises and rose to the challenges, Lee Man Tat has turned a small sauce store into a huge business, leading Lee

Kum Kee to march from Macau and Hong Kong to the colossal market of Mainland China, and subsequently strode to the world. Lee Kum Kee's products even went beyond the limit of the sky into the outer space.

Lee Man Tat's talent in seeing the big from the small as well as turning crisis into opportunity had enabled him to take immediate action once he sensed possible conflicts and cracks within the family in his autumn years. His innovative step to preempt family crisis was setting up the modern institution of Family Council to resolve issues from within the family. Since its mechanism of conflict-resolution is purposely designed to address problems at root, the Family Council has proven to be an effective institution to negotiate differences for consensus. Based on such family consensus, Lee Kum Kee is able to continue on track healthily and energetically.

10

Friendships that go far beyond business partnerships: Lee Man Tat (4th right), Chen Ying Ming (3rd right), and Chow Sing Lam (1st right).

WISDOM OF A TIDE CHANGER

To be honest, I am very ambitious and have very big dreams. Many people have said to me: Lee Kum Kee has surpassed its dream in repeatedly sending its products to the outer space, what more could we wish for? However, I always believe that one should have a contented mind in life, but not in business. An entrepreneur should never be content.

Lee Man Tat

INTRODUCTION

There is always a set of business ethics behind the legend of every extraordinary entrepreneur which allows the individual to overcome difficulties and to excel. In general, these ethics reflect not only the entrepreneur's distinctive philosophy of life, ethos, and inspiration, but also their unique wisdom, vision, and humanity concerns. Lee Man Tat's life story is no exception. The practice of these business ethics, coupled with his talent of seeing the big from the small as well as grasping opportunities from crises, has enabled him to lead Lee Kum Kee in crossing barriers, one after another, to scale new heights, thus creating the legend that he is, from the perspectives of an individual, a family, and an enterprise.

Lee Man Tat's business ethics, in practice, could be summed up in four aspects. They are first and foremost, the ultimate meaning of business: Si Li Ji Ren (considering others' interests); secondly, the way of running a business: pragmatism and integrity; thirdly, the motivation to pursue a business: constant entrepreneurship; and fourthly, the human capital of a business: picking the right man for the right job. On the face of it, these elements appear to be neither extraordinary nor enigmatic, and may even sound like clichés. However, knowing such rhetoric is one thing, but putting such into consistent practice is much more difficult. Lee Man Tat not only takes these moral principles as his mottos, but has strictly followed these in his interactions with people, in running his business, and in governing his own family. In the course of practising such business ethics, he has also developed an inclusive ideology to assimilate the modern elements of business management while upholding the fine tradition of Chinese culture as the core, as well as the mind-set to always focus on major achievements with little regard to the trivialities. This reflects his wisdom and explains his ability to repeatedly turn the tide at crucial moments: either by grasping opportunities from a crisis or by taking effective measures to pre-empt a crisis.

"Si Li Ji Ren"

To Lee Man Tat, the core value of Si Li Ji Ren (considering others' interests) is simple, straightforward and obvious, as it is all about mutual benefits and the sharing of prosperity. He may not be able, like his well-educated children, to systematically and comprehensively turn the abstract narrative of Si Li Ji Ren into concrete steps of implementation, yet based on what he has learnt from the words and deeds of his father and grandfather, he generally takes Si Li Ji Ren as a concept of considering the well-being of "all of us" instead of as an individual: if he is making money, he also ensures that his business partners are profitable; and if his business partners have fallen into difficulties, he must extend his help to them as much as possible. Apart from that, he should treat his employees well, care for and respect them, allowing them to share the economic accomplishment of the company. In fact, one can sense the tradition of Confucian business ethics (Rushang) as well as the spirit of Chinese chivalry behind the ideology of Si Li Ji Ren.

Si Li Ji Ren could also be regarded as a moral concept analogous to Confucius's "What you do not wish for yourself, do not impose it on others," as the reversed way of the saying is precisely "What you do wish for yourself, impose it on others." Also, as Sima Qian pointed out, "It was all about the pursuit of self-interest that drives people in and out in such a bustling way." Our individual desire to pursue self-interest is also a universal desire shared by everyone else, thus, if we can always consider others' interests while pursuing ours', we will be successful in every endeavour. Indeed, the moral principle concerned is simple and obvious, but it is extremely difficult to be consistently put into practice as everybody is intrinsically selfish. People tend to yearn for a monopoly of interest, endeavour to maximise their interests or their family's interests as far as possible at the expense of others'.

How does Lee Man Tat practise Si Li Ji Ren in the business arena? It could well be seen as an inconspicuous practice of Lee Man Tat in financial management. Eddy Lee mentioned that his father once guided him on the right way to review the financial statements of Lee Kum Kee, which was to check its Accounts Payable first, rather than giving priority to the Accounts Receivable like many others did. In other words, Lee Man Tat wanted Eddy Lee to check if Lee Kum Kee had any due payment to its suppliers and, if so, "pay them first". Lee Man Tat insisted that the company should maintain credibility to its suppliers and avoid any delay in payment, so as to enhance its cooperation with them. In return, the suppliers will give Lee Kum Kee priority in supply and also with quality goods. Otherwise, even if they have quality goods to offer, in the words of Eddy Lee, "They will not benefit you".

A private conversation on December 17, 2016, is worth noting as testimony of Lee Kum Kee's genuine concern over the interest of its suppliers. It was the day of the opening ceremony of the Lee Kum Kee Building in Shanghai, and guests were invited to a buffet lunch after the ceremony. A manager from a bottle cap company shared with us in casual conversation that Lee Kum Kee was willing to offer reasonable profits to suppliers, unlike other companies which always resorted to the rationale of scale economy to pressure suppliers for lower prices. The manager categorically said that "Whenever our company has quality products, we always give priority to Lee Kum Kee first".

Another unique corporate culture of Lee Kum Kee is its preference to maintain stable and long-term cooperation with its distributors and suppliers. A "long-term" relationship here refers to cross-generational collaboration that spans over decades, or even half a century, the longer the sturdier, as the saying goes. The sense of comradeship of Lee Kum Sheung and Lee Shiu Nan towards Lee Kum Kee's distributors and suppliers has been fully inherited by Lee Man Tat; in fact, he has further strengthened the comradery. The setback that Lee Man

Tat experienced in the 1950s due to his poor judgment of character prompted him to pay more attention to the personal relationship with distributors and suppliers, observing their characters and integrity through continual mutual interactions. Simply put, once the trust was established, it would be unshakable.

Harry Yeung, Senior Vice President of the Lee Kum Kee Health Products Group, joined Lee Kum Kee in 1994. Shortly after making personal contact with Lee Man Tat, Harry Yeung realised that Lee Man Tat "was not just doing business but running an 'ecosphere of friends.'" Lee Man Tat tends to value friends —in the broad sense inclusive of his long-term business associates —as well as the management staff of the company. Many stakeholders would mingle in such a circle of friends, and Lee Man Tat's unique charisma — Harry Yeung called it "Qi field" ("gathering of like minds") — that would keep him at the heart of this ecosphere of friends, thereby powerful enough to unite and influence everybody in the ecosphere.

To the gregarious and outgoing Lee Man Tat, maintaining such an ecosphere of friends is not   just a task to entertain people, but a passionate engagement with friends. Harry Yeung grinned as he commented that Lee Man Tat's ability to socialise is a rare talent, which probably could not be fully inherited by his children. Here is an anecdote that is worth noting that came from Pao Pui Li as she recalled that during the 1990s, Lee Man Tat once organised a tour with friends to the Mediterranean, but he suddenly felt sick and was advised by his doctor to rest instead of travelling. On that occasion, the doctor asked Lee Man Tat: "Which is more important: your friend or your health?" Lee Man Tat replied that his friends were more important. He then decided to head for the Mediterranean before seeking treatment in the United States.

A lot of stories could be told regarding the comradeship and deep trust established between Lee Man Tat and his business partners. Chow Sing Lam's Nansen Trading Company (HK) Ltd. was a raw-material supplier to Lee Kum Kee in the early days, and later transformed itself into a local distributor of Lee Kum Kee in Hong Kong. However, shortly after the transformation when its business had begun to move forward, a financial turmoil in Hong Kong in 1979(1) seriously shook its foundation. The two other shareholders of Nansen had already suffered a great deal and successively withdrew from the company. As Chow Sing Lam could hardly sustain his business alone, the company was in a very precarious situation.

At that critical moment, Lee Man Tat lent HK$1 million in cash to Chow Sing Lam even before he asked for help. The loan was provided for his financing over one year, and it was mortgage-free and interest-free, as this was a huge sum of money in 1979. It is particularly worth noting that Lee Man Tat was at that time far from being a tycoon in Hong Kong, and the loan interest of the market in 1979 once reached a staggering 20%. Chow Sing Lam recalled the day when he first took a flight with Lee Man Tat to South Korea for the inspection of oyster extracts there in 1973, and Lee Man Tat wrote him a note of "Si Li Ji Ren". The true meaning of Si Li Ji Ren suddenly became clear to him. After the incident, Lee Kum Kee and Nansen not only sustained their collaboration for nearly half a century, but also Lee Man Tat became one of Chow Sing Lam's best friends.

Other long-term distributors who have similar experiences of comradeship with Lee Kum Kee include the family of Cho Loin Chiu in Mexico, and the family of Chen Ying Ming in Japan. The Lee family, based in Hong Kong, is geographically far from the Chen family and very remote from the Cho family, yet they are both very close to each other beyond their business partnerships, and their friendships have spanned over three generations. Cho Loin Chiu's father, Cho Yu Chiu, became Lee Kum Kee's distributor in Mexico as early as 1918,

Lee Man Tat (1st right) and Chow Sing Lam (2nd right) at a railway station platform during their trip to study the markets in Japan in the 1980s.

which was precisely a century ago. As for the Chen family in Japan, its partnership with Lee Kum Kee has also spanned over six decades. The Daiei Trading Company was founded by Chen Ying Ming's father, Chen Wing Shun, and it was he who took the initiative to be a distributor of Lee Kum Kee's oyster sauce in Japan.

Lee Kum Kee's oyster sauce could be found in Japan as early as the 1930s, but at that time, Lee Kum Kee had not, as yet, established its sales channels there. Its products came with individual sailors who brought them from Hong Kong to Yokohama, and subsequently sold at the wharf to chefs from the Chinese restaurants in the local Chinatown. In the 1950s, Chen Wing Shun, who came from Taiwan, was keenly aware of the market potential of Lee Kum Kee's oyster sauce while doing business in the vicinity of Yokohama Wharf. He went to Hong Kong to meet Lee Shiu Nan and offered to be an agent of Lee Kum Kee's products in the Japanese market. This marked the beginning of a profitable partnership as well as a strong friendship between the two families. Chen Ying Ming pointed out in an interview that today, Lee Kum Kee's products have accounted for 90% of all the commodities distributed by Daiei. What is more glaring is the fact that over 95% of the Chinese restaurants in Yokohama's Chinatown, and over 80% of all the high-end Chinese restaurants in Japan, have chosen Lee Kum Kee's oyster sauce. As an extra effort to promote Lee Kum Kee, Daiei also adopted Lee Kum Kee's sauces to cook dishes in its own Chinese restaurant located in Yokohama's Chinatown.

Chen Ying Ming and his son Junichi Hoshino's dedication to promote Lee Kum Kee's products in the Japanese market was not merely driven by business interests, but the emotional touch triggered by Lee Man Tat. In 1994, due to heavy losses in a grand investment plan, Daiei fell into financial difficulty and was on the brink of bankruptcy, which had greatly interrupted Lee Kum Kee's sales in the Japanese market. But instead of replacing the distributor, Lee Man Tat flew to Japan to study

Lee Man Tat (2nd left) joined Cho Loin Chiu (1st left) and others in wearing sombreros and riding in a "horse and cart" while surveying Lee Kum Kee's business in Mexico.

the situation and to offer his help. Recalling this precarious moment in his life, Chen Ying Ming stressed that he was moved not because of any financial assistance from Lee Man Tat, but by his decision to keep Daiei as Lee Kum Kee's distributor in Japan even though its commercial interests had been adversely affected. Apparently, Lee Man Tat took the survival of Daiei as his first priority when considering the case. In Chen Ying Ming's own words, Lee Man Tat "Never ever abandons Daiei."

Apart from the distributors, there are also suppliers and their families who have collaborated with Lee Kum Kee for more than half a century. The Chiu Family of the Sheck Wah Tong Printing Press, and the Ho Family of the Yick Fung Printing Company, are two examples in this regard. Sheck Wah Tong has a long history and reputation in Hong Kong's printing industry. It was founded by Chiu Fu Pui in Sheung Wan, Hong Kong Island, in 1911. Chiu Sin Ching, the third generation in charge of Sheck Wah Tong, proudly shared with us in an interview that Lee Kum Kee's historic first product label was from Sheck Wah Tong and was handled by his grandfather Chiu Fu Pui. Sheck Wah Tong was then located on Gough Street, Central;(2) and the technology concerned was the traditional plate printing of lithography. The iconic label of Lee Kum Kee's Premium Oyster Sauce that depicted a "cockboat maiden" was also from Sheck Wah Tong, and was both painted and printed by the company.

Lee Man Tat was rather fastidious about the label printing and demanded good quality for a prominent image. However, in those early days, he simply relied on local firms to design and draw the product labels of Lee Kum Kee without considering any help from a professional design company. In 1960, Lee Man Tat personally designed a new and quaint brand for Lee Kum Kee. He has become an old friend of Chiu Sin Ching due to their years of cooperation, and they often gather in the Luk Yu Tea House for meals and Mahjong games since the 1980s. Sheck Wah Tong is still printing product labels for Lee Kum Kee today,

Lee Man Tat and Chen Ying Ming: a life-long friendship that has survived the harsh test of the business world, and entrenched through mutual support.

but it has already adopted the most advanced printing machine, which is truly a case of "co-prosperity" with Lee Kum Kee. Chiu Sin Ching is very familiar with Lee Man Tat's fastidious demands for quality, thus, when a tiny flaw was once found on a specific product label of Lee Kum Kee due to negligence in the factory, Chiu Sin Ching simply destroyed all of those without a single piece left. His conviction for quality is apparently influenced by Lee Man Tat, who always insists "quality is the lifeblood" of products. This is a very good example of "One takes on the attributes of one's associates."

The comradeship between Lee Man Tat and the Ho Family of the Yick Fung Printing Company is also worth mentioning. Cliff Ho was handling the design and advertising for Lee Kum Kee in the 1970s. Lee Man Tat appreciated Cliff Ho's honesty and trustworthiness, thus, he encouraged him to open a printing company to undertake part of the printing orders of Lee Kum Kee. Cliff Ho and his wife therefore decided to set up their own company, and today, almost half a century later, Yick Fung Printing Company remains a major supplier of Lee Kum Kee. It is now run by the second generation of the Ho Family, mainly responsible for the printing of Lee Kum Kee's promotional journal — *Corporate Link*. As revealed by the couple in an interview, an intriguing aspect of their collaboration is that throughout the years, they still make deals according to the traditional way of trust and good faith and have never signed a contract. Indeed, in the full ramification of mutual trust, their word is as good as their bond.

For those who once helped Lee Man Tat in his business ventures, he always keeps them in mind and tries to reward them. Hui Hung was a close associate of Lee Man Tat when he was dabbling in real estate investment in the 1960s; unfortunately, he died in 1974. From then on, Lee Man Tat never forgets to send Hui Hung's widow greetings and gifts on every Chinese New Year and major festivals. He also pays special attention to his youngest brother, Hui Chi Hing, as he particularly highlighted such an attribute of Lee Man Tat in caring about his business associates and staff in an interview, stressing that it is by no means a common deed shared by wealthy people. When Lee Man Tat took the helm of Lee Kum Kee in the 1970s, Wong Kwok Hei assisted him in exporting products to the United States, and took care of Lee Kum Kee's business on the West Coast of the United States, and he later died there. Chang Chin Vai, a long-serving staff member of Lee Kum Kee, noted that a company notice related to Wong Kwok Hei was issued one day, which stated that Lee Man Tat would like colleagues "To mourn for Mr. Wong for 15 minutes" on a specific day (being the anniversary of Wong Kwok Hei's death).

In the 1980s, following the new policy of reform and the opening-up in Mainland China, Han Changde was sent to Hong Kong as a trade representative of Liaoning Province, actively promoting Liaoning's business and investment opportunities in Hong Kong. He introduced and arranged trips for Lee Man Tat and Eddy Lee to Liaoning Province, with the main purpose of inspecting oyster extracts and the mariculture industry there. Han Changde later also assisted Lee Kum Kee to set up plants in Liaoning Province, including a seafood processing factory as well as those which deal with raw materials such as chili. Eddy Lee mentioned that later, when his parents visited Northeast China again in the late 2000s, they braved the extremely cold winter of Liaoning to make their way to a remote place in the Changbai Mountain area, just to pay a visit to the retired Han Changde in his hometown, and he was greatly moved by their warm gesture.

Likewise, Chen Jingmiao greatly facilitated Lee Man Tat's investment in Fuzhou in the 1980s. Lee Kum Kee built a factory there later, but at the beginning, it was Chen Jingmiao who guided Lee Man Tat to Fuzhou. Chen Jingmiao was then the Vice Chairman of Hua Rong Limited, a liaison office of the Fuzhou City Government in Hong Kong.(3) After his death, his son Chen Rui once shared with friends that on the eve of every Chinese New Year, Lee Man Tat would send Lee Kum Kee's staff in Fuzhou to his home to offer greetings, and to present Chen Jingmiao's widow with a laisee (red envelope) for the New Year. Also, when Lee Man Tat decided to establish a production base in his ancestral hometown of Qibao, Xinhui, in the 1990s, He Xiansong, then Mayor and later Party Secretary of Xinhui City, offered tremendous help to kick start the project. After his retirement, He Xiansong unfortunately suffered from a stroke in October 2012. When Lee Man Tat heard the news, he immediately travelled from Hong Kong to Xinhui to visit him.

Lee Man Tat has always been considerate to his employees, particularly to long-serving staff. Chang Chin Vai has been the only member of staff managing the old store of Lee Kum Kee in Macau since the 1970s. In the early days, he did not own a house even after getting married. Chang Chin Vai once took a long break for recuperation after major surgery, and Lee Man Tat paid him a visit in Macau. When the two strolled down the street, Lee Man Tat spotted a building under construction. He turned to Chang Chin Vai and said: "Ah Vai, this building looks pretty good. The most important thing for you to settle in Macau is to have your own shelter." Chang Chin Vai replied with a wry smile that he did want to own a house, but was short of money. Lee Man Tat then asked who the developer of the building was. Surprisingly, they found out that the project was one of his childhood friends, Ho Shiu Chung. "Can you ask him how much it is?" "It is worth several tens of thousands." "That's fine. I will support you." As a result, Lee Man Tat offered Chang Chin Vai funding of MOP$30,000 for an apartment that costed MOP$62,000, and told him, "Pay me back when you have the money; otherwise, just forget it." Chang Chin Vai smiled as he admitted that he still owed his boss the loan today.

Lam Yu, Senior Vice President of the Lee Kum Kee Health Products Group and the Managing Director of Infinitus (China) Company Ltd.,

Fervent in Sino-US trade and relations, Lee Man Tat regularly attends gatherings organised by the Sino-US Foundation. Lee Man Tat and wife with Tung Chee Hwa (former Chief Executive), Carrie Lam (incumbent Chief Executive), and Sammy Lee after one of the activities held by the Foundation at the Government House.

is also impressed by Lee Man Tat's attention to the welfare of his employees. One year, Infinitus performed particularly well, and when Lam Yu briefed the Board of Directors on their accomplishment, Lee Man Tat stressed that the contribution of the staff must be treasured. He then turned to Sammy Lee, who was in charge of Infinitus, and said: "Sammy, show them your appreciation!" It served as a reminder to Sammy Lee to better reward the staff and share the fruits of prosperity.

Lee Kum Kee used to run a long-service travel scheme to reward their long-serving staff. For staff who had served Lee Kum Kee for a decade, an overseas tour was the reward, and the staff concerned could enjoy the reward again after another five years. Take the Tai Po plant as an example, staff with 10-year seniority could travel to Malaysia, while those with 15-year seniority would head to the United States. As for those with 20-year, 25-year, and 30-year service, respectively, their spouse or another family member could also join the tour, and the destinations ranged from Japan (20-year seniority), Europe (25-year seniority), to a cruise to the United States (30-year seniority). However, as the number of long-serving staff kept increasing, an overseas tour often involved many people, thus adversely affecting the daily company operation. Lee Kum Kee has therefore changed its incentive scheme from overseas tours to awarding commemorative gold coins in recent years.

Pragmatism and Integrity

Lee Man Tat is a pragmatic man who dislikes fantasy and does not deliver empty words. Although he is fond of social gatherings with lots of his friends and is generous in giving tips, however, if he does not get value for money, he will not reciprocate. Such a pragmatic style is most vividly reflected in his obstinacy in choosing economy class while taking a flight, which is certainly not due to frugality. Tsoi Cheong Ming, Choi Cheong Too, and others all have mentioned in interviews that Lee Man Tat is a big-hearted person who gives generous tips every time he has meals in the restaurants that he frequents. Waiters and waitresses, thus, will enthusiastically flock to his table and warmly greet him. It is therefore clear that unlike the late Wang Yung-ching, the richest man in Taiwan, who kept on using a same towel for nearly 30 years, Lee Man Tat has a different rationale in his insistence on taking economy class on flights.

Lee Man Tat's rationale, in this regard, is that the business class is much more expensive than the economy class, but its service as well as seat comfort are not much different, hence, it is not worth it. According to Raymond Lai, Lee Man Tat took the economy class, as usual, on a flight to the United States in April 2008, even though he was already 79 years old and was travelling to the special occasion of his induction into the Babson College's "Academy of Distinguished Entrepreneurs." In fact, when the almost 88-year-old, Lee Man Tat attended the opening ceremony of the Lee Kum Kee Building and the naming ceremony of the Infinitus Tower in Shanghai in December 2016, he still stuck to his choice of economy class on flights despite family members and friends trying hard to persuade him otherwise.

Nevertheless, according to Vincent Ko, who often accompanied Lee Man Tat on his overseas tours, it is not difficult for Lee Man Tat to fly economy class, as he can easily fall asleep: "Once he finished his meal on the plane, Lee Man Tat fell asleep quickly; and when he woke up, he thought about Lee Kum Kee's business." Vincent Ko, however, added that Lee Man Tat did not insist on senior managers of the company to take economy class with him. An intriguing phenomenon is that Charlie Lee, currently Chairman of the Lee Kum Kee Sauce Group, follows his father's style of pragmatism. He is also prudent in spending and also takes economy class while travelling abroad.

Apart from pragmatism, Lee Man Tat attaches paramount importance to integrity, which is reflected in his determination to safeguard Lee Kum Kee's reputation and the quality of its products. In short, he strives to keep the premium quality of Lee Kum Kee's products at all costs. Lee Man Tat's conviction in raw-material procurement, as Chiu Sin Ching observed: "Since I have paid a good price, I deserve quality goods, and please reserve ALL your quality goods for me." His willingness to pay higher prices for the best is precisely due to his eagerness to ensure product quality. That was why in the 1970s, when Lee Man Tat procured oyster extracts from Japan through Chow Sing Lam, he requested them to secure all the premium oyster extracts for him whatever the price might be. Lee Man Tat's strong preference for excellent quality, in fact, resembles the Japanese. Indeed, he appreciates various aspects of the Japanese way of life and feels particularly congenial with Japanese entrepreneurs, so much so, that he once considered sending his children to Japan for their education.

On this front, Lee Man Tat's daughter, Elizabeth Mok, recalled the following:

> A long time ago, before we studied abroad (in the United States and Canada), I forgot whether it was David or Charlie who almost furthered their study in Japan, even with their school uniforms ready then. Perhaps at that time daddy thought that the Japanese were excellent in engineering for food manufacture, so he did consider sending them to Japan to study the language.

> There was a distributor in Japan with whom we had collaborated for many years. Daddy might be deeply impressed by his visit to the Japanese factories, and shared with us upon his return that places managed by the Japanese were all very clean and systematic, so he hoped I could learn about it. I think he really prefers the Japanese style of management. But eventually my younger brothers didn't go to Japan, whatever the reason might be.

...Daddy later said to me that certain Japanese packaging was rigorous, and urged me to study packaging technology in Japan. I thought I should return to the United States and learn it from the Americans, but daddy felt otherwise: apparently, he was more impressed by the exquisite aspect of the Japanese culture. However, to study in Japan one must master their language first. It was in 1979 or 1980, the Japanese were not good at speaking English. So mommy found me a Japanese language teacher, and I attended his class seven days a week... After learning for a while, the teacher said I was basically OK, but still weak on conversation, thus unable to work in Japan yet. Daddy then arranged me to study at a Japanese language school in Kobe.

He had a Taiwanese friend who also surnamed Lee and lived in Kobe... I attended Japanese language class in the morning, and went to Uncle Lee's office in the afternoon. The office was pretty small, with only six or seven employees. Daddy wanted me to learn how to trade there and practise my conversational Japanese at the same time. Most of them were female staff; however, when they knew I had returned from the United States, they were eager to learn English from me and as a result, no one spoke Japanese to me.

Elizabeth Mok later shuttled between Japan, the United States, and Hong Kong for various reasons, and her study in Japan was repeatedly interrupted. She was simply unable to concentrate on learning Japanese. Meanwhile, Lee Kum Kee's business was booming and the family needed assistance, Lee Man Tat, thus, reluctantly called her back to Hong Kong to help, leaving the mission unaccomplished.

There are many anecdotes about Lee Man Tat that relate to his meticulous attention to quality. Before the adoption of containers in the shipping industry in the 1960s, Lee Kum Kee's oyster sauce was loaded inside wooden cases to prevent any damage caused by impacts in rough seas. When the industry began to switch from wooden cases to carton boxes, Lee Man Tat was not sure as to whether these carton boxes were solid enough to stand the impacts and protect Lee Kum Kee's oyster sauce from damage and deterioration. In order to test these boxes, he conducted an experiment by carrying a carton loaded with Lee Kum Kee's oyster sauce to the fourth floor, and then pushed it down all the stairs to the ground floor, testing its firmness as well as its reliability in anti-collision.

Another anecdote related to carton boxes happened in Australia in the late 1980s. Lee Kum Kee's general distributor in Australia is the Oriental Merchant Pty. Ltd., which is likewise a family business and a long-time partner of the Lee family. Bernard Yiu, CEO of the company, recalled that in the late 1980s, when Lee Man Tat and his wife went to Australia, they visited his company's warehouse in Melbourne and astutely noted that the carton boxes designed to hold Lee Kum Kee's products were not in good shape. They were not strong enough to stand the impacts and might therefore damage the quality of the products. Lee Man Tat immediately improved their design upon his return to Hong Kong.

Lee Man Tat had to deal directly with distributors in the United States after he had dispensed with Kam Shan Chong. In the face of certain picky distributors there, Lee Man Tat reacted quickly to complaints in order to safeguard Lee Kum Kee's reputation. For instance, when one of them complained about the flavour of a batch of Lee Kum Kee's oyster sauce, though he knew it was not an issue of quality, he recalled the goods immediately and sent out a different batch to him. The transaction cost him dearly, but he believed it was certainly worthwhile for the reputation of Lee Kum Kee (Wu, 2013).

Stephen Chung, Senior Vice President of the Lee Kum Kee Health Products Group, recalled that in 1994, after he had just joined Lee Kum Kee as a Financial Manager, a group of mid-level managers were lucky enough to have the opportunity to visit the Huangpu plant in Guangzhou led by Lee Man Tat personally, which could be regarded as some form of "orientation." Stephen Chung was truly impressed by two incidents that day. The first was when Lee Man Tat saw a group of female workers peeling garlic in the Huangpu plant. He asked the General Manager why he did not buy garlic which had already been peeled. Stephen Chung said to himself: "Wow, this boss is really something. He not only has the concept of cost-effectiveness, but is familiar with the process of sauce manufacturing as well." Later that day, since Lee Man Tat had scheduled a meeting with the management of the Huangpu Plant, he let this group of mid-level managers sit in on the meeting. There was a report on a complaint about the quality of oyster sauce which had remained unresolved for over a week. Lee Man Tat was furious after he learnt about the details of the case blasting the cadres of the plant for being dilatory. "Don't you know such a complaint will seriously hurt Lee Kum Kee's credibility? I give you two days to resolve this problem!"

Meng Suhe,(4) President of the Chinese Institute of Food Science and Technology (CIFST), met and befriended Lee Man Tat while she visited Hong Kong on a business trip in 1996. A few years later, the inspection authorities of the United Kingdom were concerned about chemicals contained in various soy sauce products from China and requested an on-site check of soy sauce productions in China. Meng Suhe admitted that at that time, the soy sauce industry of China was flooded with additives, thus dared not open up for them to check. She shared with us her reflection of the event which is worth pondering upon:

> Because we dared not let them come to China for an inspection, soy sauce products from China were blocked from export. Lee Man Tat later said to me that we could invite them to Lee Kum Kee's plant, and his team would receive them. Upon our return, we immediately reported to Li Chaowei, the then [Deputy] Director of the [Bureau

of Import and Export Food Safety,] General Administration of Quality Supervision, Inspection and Quarantine that Lee Kum Kee welcomed the inspection. The authorities of the United Kingdom said if we could let them inspect one factory, they could then talk about opening up of their market to our soy sauce products with us. They eventually visited Lee Kum Kee's plant and carried out rigorous inspections there. Consequently, the British market was opened up....

Thus as we can see, quality control is always a conviction of Lee Kum Kee. Lee Kum Kee was among the first few enterprises to go abroad. When its products entered the Japanese and the American markets, it was facing the most stringent food-safety standards. At that time, there were very few exports of soy sauce and oyster sauce products from other Chinese manufacturers. Whoever became the pioneer in exploring international sauce markets was also the first to be subjected to international standards. These have facilitated Lee Kum Kee to adopt international standards on food safety much earlier than other enterprises.

Meng Suhe reckoned that Lee Kum Kee's food safety control was the best in China's food industry today. She pointed out that "Lee Kum Kee does not enjoy the lion's share of the Chinese market, but it must be the best in quality." To her, food safety encompasses three aspects, namely, ethics, regulation and technology: "If the core element of ethics is absent, technology will become a mere accomplice of fraudulence."

Nevertheless, quality regulation and inspection on food products involve various technicalities, and hence it is difficult for any company to meet all the requirements all the time. Indeed Lee Man Tat is always meticulous in maintaining the highest standard of Lee Kum Kee's products as the core heritage of the Lee family, yet unfortunately, Lee Kum Kee was hit by an unexpected quality crisis between 1999 and 2001. In late 1999, a report published by the British Food Standards

Authority (FSA) noted that 3-MCPD(5) was found in many soy sauce brands in the market, including Lee Kum Kee's soy sauce.

Before this incident, the European Commission had never announced any 3-MCPD standards; the FSA Report made a huge negative impact on Lee Kum Kee's image. Eddy Lee, then Chairman of the Sauce Group, immediately spent US$ 260,000 to improve the manufacturing processes, so that 3-MCPD was reduced to non-detectable level. But a bigger blow to Lee Kum Kee occurred when the FSA conducted another survey in 2001 and discovered that the amount of 3-MCPD in 22 brands of soy sauce, including five owned by Lee Kum Kee, still exceeded the standards set by the European Commission on 8 March 2001. Needless to say, it deepened the crisis of confidence amongst Lee Kum Kee's customers; even the markets in Australia, Singapore, Egypt and South Africa began to pay attention to the case. However, as a matter of fact, FSA had used the previous batch of Lee Kum Kee's products as samples, which was produced before Lee Kum Kee adjusted its manufacturing processes and the European Commission announced the 3-MCPD standards.

At that point, Lee Man Tat and his sons were all highly conscious of the crisis. Lee Man Tat entrusted Eddy Lee to deal with the crisis, and in retrospect, the strongest impression that Eddy Lee had was: "I was seriously short of sleep for around two months, as I had to constantly cope with the influx of enquiries worldwide: from the testing institutions, customers, media, and employees." After the issue of the FSA Report, Eddy Lee was aware of the importance of 3-MCPD testing. Therefore, right after Lee Kum Kee's testing laboratory had been accredited in January 2001 as the first testing laboratory in Hong Kong to conduct 3-MCPD testing, it presented laboratory certificates to prove that its new products no longer contained 3-MCPD. The FSA also issued a statement to clear its name, but the whole saga had inflicted damage on Lee Kum Kee's reputation. Although it actively set up hotlines for consumers, hired a public relations company to clarify via advertisements, and presented endorsements from government food authorities around the world, its image of premium quality had suffered throughout the event (Yim & Mak, 2005: 67-68).

From then on, Lee Kum Kee set an unequivocal rule of testing that applies to any product of the company, which is, to follow the most stringent testing standard in the world in all cases. Summing up the lessons, Lee Man Tat and Eddy Lee transformed their conviction on quality into a concise and powerful slogan, namely "100 - 1 = 0". A few years later, in 2005, the outbreak of the Sudan dyes incident in China deeply worried both the locals and Hong Kong residents.(6) Michael Sze, former Executive Director of the Hong Kong Trade Development Council (1996-2004), and currently an Advisor to the Board of Directors of Lee Kum Kee Company Ltd., was on a hike with Eddy Lee one day during that period of time, and found Eddy Lee in a very relaxed mood. He curiously asked him about his composure in the face of the Sudan dyes crisis, and he confidently replied, "Rest assured, Lee Kum Kee will be fine." This was because all the chili that Lee Kum Kee procured came from cooperative farming with reliable farmers; and more importantly, Lee Kum Kee strictly tested their harvest, thus eliminating any risk of illegal additives.

Constant Entrepreneurship

Lee Man Tat is a commander filled with entrepreneurial passion and relentlessly strives for his causes. He has the courage to take risks and is innovative, thus drives the constant progress of Lee Kum Kee. The embodiment of his pragmatism in entrepreneurial activities is to focus on results rather than means. Ho O Sang, a long-serving Lee Kum Kee staff member, observed that Lee Man Tat is free of ideological baggage — be it a Chinese-style, a Western-style, or a Japanese-style approach, so long as it could help to make deals and facilitate the business, it is considered a good approach.

Lee Man Tat enjoys playing mahjong very much, and his gaming style on the mahjong table also reflects such an aggressive spirit of entrepreneurship. When playing, Lee Man Tat always finds himself comfortable in any venue without any taboos; it could be somewhere inside the International Funeral Parlour managed by his best friend Leung Chi Fung, or the elegant Tai Tung Restaurant, or the Luk Yu Tea House located in Sheung Wan and Central, (7) respectively, or his "Panda" private yacht anchored in a quiet location. According to Loi Kam Sheung and his wife, who are business partners and old friends of Lee Man Tat, he enjoyed playing mahjong with his friends on his yacht; and when everybody became tired, he might steer the yacht to a deeper bay and jump into the sea for a swim with them.

Lee Man Tat often advises his friends and subordinates that they should not think about the chances of winning or losing first before becoming engaged on something different. According to He Yongxiang, a long-serving employee of Guangzhou Infinitus, Lee Man Tat regularly cautioned them at the mahjong table that "If you think about winning or losing first, you will become hesitant and indecisive, thus not being able to achieve something big." But these words may only reveal half of his true colours. What Lee Man Tat implied in such a remark was that one must have the courage of adventure to make decisions and take risks in the spirit of "6677" which he so vigorously promotes. It suggests that if one has 60% to 70% confidence to succeed in something, and also believes that this is on the right course, one should proceed without any hesitation or procrastination. Lee Man Tat not only loves to share with others this 6677 Spirit, he also practises it accordingly in his business ventures. The founding of Nanfang Lee Kum Kee was the best example in this regard.

However, even though Lee Man Tat could resist thinking about winning or losing before engaging himself with something, once started, his mind-set will turn into the "must-win" mode. In the eyes of his children and close associates, Lee Man Tat is not only extremely aggressive and full of entrepreneurial spirit, he is also eager to excel others in all his pursuits. The car-racing story of Lee Man Tat with the provocative guy in Chapter 5 is one of the best examples in this regard. But even when leisurely playing mahjong, according to Charlie Lee, Lee Man Tat is also "a masterly mahjong player", and once he joins the game, there is no question of father and son but only of winner and loser. He will strive hard to win! He always says that once we step into any industry, we must achieve leadership status, and we must win. It is like playing football, don't join the game if you don't want to win... (*Yi Zhoukan*, 2014: 79).

Lee Kum Kee is the "main business" of Lee Man Tat, but he has much more than that. Since the 1950s, Lee Man Tat has dabbled in innumerable "subsidiary businesses." Almost all his friends and relatives agree that he is always thinking about business and incessantly searching for business opportunities, and will try them out if there is a chance. In addition to the businesses of real estate, coffin wood, and restaurants in his early days, when the newly constructed Tin Wan plant was ready in the early 1980s, Lee Man Tat and Eddy Lee made a special trip to Germany to find equipment needed for mechanisation of the factory's production. However, "Once they arrived, Lee Man Tat saw business opportunities everywhere. When he spotted the waffle-maker, he nearly bought it to make waffles back home. Later, he discovered a machine that made spring roll wrappers, and he wanted to explore that business as well... He always wanted to explore a field whenever there appeared to be business opportunity," as Eddy Lee amusingly recalled their trip to Germany.

Lee Man Tat's businesses went beyond those. In the 1980s, while in the United States, he also worked with friends to set up the Lee's Company, which dealt with tableware, porcelain bowls, and other products. Some years ago, when he gained some experience in breeding Japanese koi carp as a pastime, he passionately collaborated with his good friend Raymond Lai and turned this hobby into a business. These "subsidiary businesses" of Lee Man Tat are usually rather successful and generate some profits, but his engagement in these ventures may not necessarily be motivated by monetary gain, but by a natural impulse of entrepreneurs. To be more specific, it is a classic entrepreneur that actually enjoys the process of starting a business as a hobby (Schumpeter, 1934).

Lee Man Tat's entrepreneurial spirit could also be seen in his astuteness and innovation in developing sauce products. The XO sauce, which is commonly seen in Hong Kong today, did not exist before 1992. The prototype of the XO sauce came from the Spring Moon Cantonese restaurant at the Peninsula Hotel. Elizabeth Mok was impressed by the delicacy of this appetiser, and suggested to her father to try it. Lee Man Tat found it very delectable on tasting and asked the research staff to develop a similar product. In 1992, Lee Kum Kee launched its XO Sauce, with dried scallop, Chinese ham, and dried shrimp as its main ingredients. The product soon became widely popular in the market, and prompted many companies in the industry to imitate.

The process of creating Lee Kum Kee's "Seasoned Soy Sauce for Seafood" was rather similar. When Lee Man Tat tasted a very delicious steamed fish for dinner at a country club, he immediately asked the chef for the know-how to make the sauce. He instantly asked the research staff to develop this sauce upon his return to the company. When they produced a batch of samples, Lee Man Tat sent them to the company's directors, as well as his relatives and friends, for them to try. Michael Sze was also on the list of his "tasting team" and frankly admitted that the proto-type of the sauce did not taste well. But after repeating the experiments, by 1997, Lee Kum Kee finally launched its Seasoned

Soy Sauce for Seafood, which remains a highly popular Lee Kum Kee product today (Sun, 2014). Michael Sze disclosed in an interview that this sauce alone had reached an annual turnover of over HK$1 billion in Mainland China.

Rome was not built in a day. Lee Man Tat's perseverance in various attempts is the key to Lee Kum Kee's global reputation and distinguished achievements today. His entrepreneurial spirit stems from a mentality of never being complacent of the status quo, as well as the enjoyment over the pleasure, recognition, and the meaning of life derived from this process of struggle. It propels his continuous advancement in every moment of his life. For such an aggressive entrepreneur, Lee Man Tat's own words are the best elaboration for his motivation:

> To be honest, I am very ambitious and have very big dreams. Many people have said to me: Lee Kum Kee has surpassed its dream in repeatedly sending its products to the outer space, what more could we wish for? However, I always believe that one should have a contented mind in life, but not in business. An entrepreneur should never be content.

Picking the Right Person for the Right Job

A large enterprise cannot function without the collective toil and sweat of its employees. Lee Kum Kee's continuous development and success could also be attributed to Lee Man Tat's ability to pick the right person for the right job, while allowing individuals to show their full potential in making contributions to the company. In fact, Lee Man Tat prefers to invite his close relatives and friends to become staff members of Lee Kum Kee, but the choice is limited only to the reliable and talented ones.

As a young and inexperienced businessman in the 1950s, Lee Man Tat trusted the wrong person and suffered both emotionally and financially. He learnt the lesson and realised that the true colour of a man needed time to be observed. Since then, he would pay attention to every detail of a person's behaviour while interacting with them, and he began to recruit friends and relatives whom he was familiar with to join Lee Kum Kee. As his considerations on the company's human resources include not only the candidates' ability but also their integrity and reliability, employing friends and relatives could spare him the nuisance of character assessment. He invited brothers-in-law from both his family and Choi May Ling's family to join Lee Kum Kee only when they had achieved successful careers in their respective fields and shown their talent, and he would put them in positions commensurate with their abilities.

Lee Man Tat always makes good use of the family networks to look for talent. On the side of the Lee family, there are Lee Man Tat's brothers-in-law Ho Wah Yim, Tang Fook Chuen, and Lo Woon Fai, who served the company. Ho Wah Yim was in charge of Lee Kum Kee's general administration and procurement for years, and had contributed significantly to the construction of both Tin Wan and Tai Po plants. Tang Fook Chuen, currently an Advisor to the Sauce Group Chairman, graduated from the Grantham College of Education and furthered his study abroad in Canada. He was invited to join Lee Kum Kee when it moved from Wong Chuk Hang to Tin Wan in 1982. At that time, Tang Fook Chuen was already a prominent educator that specialised in special education and had successfully served as principal of a special school. Throughout his career in Lee Kum Kee, Tang Fook Chuen had retired three times, yet continually re-joined on the behest of Lee Man Tat. Tang Fook Chuen's contribution to Lee Kum Kee is remarkable. Sammy Lee particularly mentioned his "super rescue task" in 1998 when Nanfang Lee Kum Kee (i.e. today's Infinitus) was successively hit by two major crises, and Tang Fook Chuen was sent to Guangzhou for several years, with the mission to streamline Nanfang Lee Kum Kee's administration and curb its expenditure. He finally helped to save the company from its fall. As for Lo Woon Fai, he had worked for a construction company prior to joining Lee Kum Kee and was invited by Lee Man Tat to manage its procurement department. He later took charge of the construction of the Xinhui plant as well as the procurement contracts.

As regards the Choi family, in addition to Choi May Ling's brothers Choi Cheung Yiu, Choi Cheung Woon, Tsoi Cheong Ming, and Choi Cheong Too previously mentioned, her brother-in-law Lee Tat Wing had also served the company. Eddy Lee acclaimed Lee Tat Wing for his excellent command of English, adding that he had served in the sales and marketing department of a British company for years, and was thus highly experienced in sales and networking. Lee Man Tat netted him into Lee Kum Kee for negotiations with foreigners and handling English communications, so as to remedy a deficiency in the skill-set of the company management.

In fact, the Lee family and the Choi family have always maintained very close interactions. When Lee Man Tat's children were studying in primary schools in Hong Kong, they would spend part of their summer vacation at the Choi's mansion in Macau, whereas Lee Man Tat only occasionally visited his in-laws. Choi Hac Teng later sold the mansion and migrated from Macau to Kowloon, thus facilitating even more frequent contacts between the two families. On weekends, Lee Man Tat often took the ferry with his family from Hong Kong Island to Kowloon to visit the Choi's home. Members of the Choi family often flocked to Choi Hac Teng's home during the weekends in their longstanding tradition of happily enjoying games of mahjong.

Also worth mentioning is the Choi May Ling alumni network of Lingnan University. It was through this network that Lee Man Tat met Shi Yushen, who helped him to explore the condiment market of Guangzhou in the 1980s. The network likewise connected Lee Man Tat to Yuan Zhenyuan, who was then a Professor at the South China University of Technology, and a highly experienced and respectable expert in the condiment industry of China. Yuan Zhenyuan and Yang Jieming, another expert recommended by the former to Lee Man Tat, were the two key figures that helped Lee Kum Kee to finally produce its own soy sauce in the 1990s.

Apart from friends and relatives, what kind of people does Lee Man Tat prefer? Obviously, they are those with honesty and integrity. According to Cliff Ho, in the 1970s, when he was still handling the design and advertising business for Lee Kum Kee, Lee Man Tat said to him one day: "You are not shrewd enough, and that suits me." He then encouraged Cliff Ho to set up his Yick Fung Printing Company, and entrusted it as Lee Kum Kee's printing supplier. It shows that the "Smart Alecs" of this world may not be reliable in Lee Man Tat's eyes. He would rather shun these people for those whose loyalty is beyond reproach.

Conclusion

A trademark infringement lawsuit filed by Lee Kum Kee in Malaysia could well reflect and sum up Lee Man Tat's business ethics and his wisdom. During the 1980s, Lee Kum Kee's turnover in Malaysia suddenly plummeted. Lee Man Tat entrusted Eddy Lee and Tang Fook Chuen to go to Malaysia to investigate, and as a result, they found counterfeit products. A local factory was counterfeiting the Panda Brand Oyster Sauce with a duplicate resemblance of both its trademark and packaging, and they were sold in 18 local supermarkets. Allowing inferior oyster sauce to mar his Panda Brand was something unbearable to Lee Man Tat; thus, in 1984, Lee Kum Kee filed a lawsuit against the factory and the supermarkets even though it was expected to be costly both in terms of time and money to spend. The case was handled by

Samuel Yu, a lawyer from the Deacons, and supervised by Tang Fook Chuen. The team also sought legal support from Malaysian lawyers. After years of litigation, Lee Kum Kee finally won the case. However, as expected, regardless of the judgment, Lee Kum Kee had incurred a high cost which certainly exceeded the compensation that it could recover. The local factory had closed down by then, but the 18 local supermarkets still bore the responsibility to compensate Lee Kum Kee.

It did not take long for Lee Man Tat to decide not to proceed with the claims. His objective to defend the brand had been achieved, and in taking a broader perspective of the situation, he had no desire to force bankruptcy on these supermarkets. Therefore, instead of pursuing reparations, he suggested that they use the money for the sales promotion of Lee Kum Kee's products in their supermarkets. According to Tang Fook Chuen and Samuel Yu, the owners of these supermarkets were so grateful to Lee Man Tat's unexpected move that they actively promoted Lee Kum Kee's products, pushing up the sales of the Panda Brand Oyster Sauce in Malaysia. A change of mind-set and perspective based on Si Li Ji Ren had turned adversary in court to partners in business. In 1997, Lee Kum Kee even decided to build a plant in Kuala Lumpur, Malaysia, for the production of halal oyster sauce in the region.

In sum, Lee Man Tat's distinctive business ethics and his individual style in running the business and family reflect his immense wisdom. Such wisdom closely associated with his peculiar ability in seeing the big from the small, or seeing the universe from a grain of sand. This allows him to spot something meaningful and momentous in triviality, which is usually ignored by the ordinary people. He also fully understands that losses could be gains, and that breakthroughs could come from disadvantaged scenarios, thus enabling him to make much better choices.

The four aspects of his business ethics — namely Si Li Ji Ren, pragmatism and integrity, constant entrepreneurship, as well as picking the right man for the right job — have traces in traditional Chinese culture, but some are exclusively "Lee Man Tat." These practices, coupled with his talent in seeing the big from the small, have allowed him to seize opportunities when in crises, and subsequently changed the tides. Lee Man Tat is, indeed, truly a legend of his time.

1   It was believed to be caused by a sudden visit of the former Governor of Hong Kong Murray MacLehose to Beijing, in which he learnt that the Chinese government was adamant on resuming the exercise of Hong Kong's sovereignty in 1997.

2   Sheck Wah Tong Printing Press moved to King's Road, North Point, Hong Kong Island, in 1964.

3   Chen Jingmiao was a former teacher and former Principal of Fuzhou No.3 Middle School, and later became Deputy Mayor of Fuzhou City. He had been stationed in Hong Kong since April 1985, and was transferred back to Mainland China in 1992.

4   Meng Suhe has become a Consultant to the Expert Committee on Food Safety and Nutrition of Lee Kum Kee since 1999.

5   3-MCPD: 3-monochloropropane-1,2-diol.

6   Sudan dyes are industrial dyes not permitted to be used as food additives. However, on March 4, 2005, Heinz's chili sauce produced in Mainland China was detected as Sudan I in Beijing. Within one month, Sudan I had also been discovered in food items of Kentucky Fried Chicken and many other restaurants, as well as in various food products of Mainland China. It became a sensational incident then.

7   Since he settled on Hong Kong Island in the 1950s, Lee Man Tat prefers to have his meals and mahjong games with his friends and business partners in the Tai Tung Restaurant and the Luk Yu Tea House. The Tai Tung Restaurant opened in 1925, but was wound up many years ago. It was located at the junction of Des Voeux Road Central and Mandarin Lane in Sheung Wan, just opposite today's Wing On Centre. Luk Yu Tea House started its business in 1933 on Wing Kut Street, Central, and has moved to the present site on Stanley Street since 1976.

PROLOGUE

WHY LEE MAN TAT OF ALL PEOPLE?

When we met Lee Man Tat for our first in-depth interview with him, he welcomed us with a humble remark: "I'm just an ordinary businessman, nothing extraordinary to write on". Indeed, it was a self-depreciating remark, but it did touch on an issue worth pondering about: Why do we want to write about Lee Man Tat? Questions that arise from this issue are: What kind of distinctive historical and social phenomena could be reflected in his life? What kind of lessons, experience, and wisdom could he leave behind for society? And, in terms of personality, what is the uniqueness of Lee Man Tat that is worth highlighting?

Lee Man Tat and his family have always kept a low profile in their public lives. Thus, compared to other prominent and glamorous families in the business world of Hong Kong, Lee Man Tat and the Lee family are relatively less captivating, and therefore seemingly "not a great deal to talk about." Likewise, the heart of Lee Man Tat's business is Lee Kum Kee, which sells sauces and health products. These are commodities that many may regard as "insignificant," hence, once again, seemingly "not worth making a meal out of it". Yet in fact, Lee Man Tat is a man full of alluring life stories and thought-provoking anecdotes, with the ups and downs of his career closely associated with the contemporary socio-political changes that have shaped the trajectory of our modern history. In this sense, Lee Man Tat is definitely a prominent figure, and one that is worth the subject of serious literary study.

To be more specific, we believe that at least three aspects of Lee Man Tat's life justify the necessity and significance for us to write about his life story:

(1) Lee Man Tat has a peculiar ability and insight of visualizing the big from the small, and his life experience has reflected the high value of such an attribute. His talent to spot and grasp opportunities from adversities is precisely because he could see the big picture from the small detail, thus enabling him to move even further in times of crises

and challenges.

(2) Lee Man Tat's perseverance and tenacity in times of hardship are truly remarkable. His persistent efforts have enabled him to lead the family business of Lee Kum Kee from small to mega and earn a place in the international business arena. This is no doubt an exceptional accomplishment, as Chinese family businesses have always been labelled as intrinsically problematic, and therefore, rarely able to achieve anything significant.

(3) Lee Man Tat's career generally reflects efforts and contributions that the Chinese in Hong Kong, Macau, and overseas communities have made throughout the history of modern China. They have particularly witnessed the tortuous course of China's development from weak to strong, as well as the rejuvenation of the Chinese nation.

These three aspects of Lee Man Tat's life have not only earned him immense respect and reputation throughout society, but, at the same time, reflected the tremendous contributions of the Chinese communities in Hong Kong, Macau, and indeed, worldwide, towards the development of modern China: such roles in this regard should not be underestimated.

The Chinese philosopher Han Feizi once said: "The sage can see how things will develop from their minute inception; the sage can also see through the end of the event from a small clue at the beginning". As mentioned in various chapters of this book, Lee Man Tat has repeatedly ridden through the storms to achieve major breakthroughs in his life and business because he has this profound ability, thus, equipping him to see many opportunities, and being undeterred by prevalent difficulties whatever they may be. Lee Man Tat could also be immune from the distraction of specific socio-political chaos, and come to grips with the key elements of development for the more effective response and planning.

Intriguingly, clues for such an ability of Lee Man Tat could be traced through the lines of the four scrolls of calligraphy presented by Liu Tai-shi to him in the 1960s. These scrolls contain not only the most inspiring phrase of "Si Li Ji Ren" to him, but also the sentence of "for a person of great ability, it takes time to finally reveal his or her true uniqueness". It could be seen as an indirect reference to the ability of seeing the big from the small. It describes a situation where people with great intelligence and ability may have some innate traits that may not initially "catch the eye" but function like the DNA of human genes that will eventually determine one's trajectory of development. This may reflect a personal feeling by Liu Tai-shi after close observation of Lee Man Tat, yet unwittingly touched on the core.

Lee Man Tat's ability allows him to stay calm and serene in front of difficulties and challenges, and subsequently grasp the right opportunity in crises to change the tides. Such a remarkable personal trait is behind the crisis management philosophy that is well worth, for many of us, to take lessons from.

As noted in *Zhong Yong (Doctrine of the Mean)* of *Liji (Book of Rites)*, "One that wants to go high must start from the bottom; one that wants to go far must start from close by". Lee Man Tat is a pragmatic and hardworking man who dislikes fancy ideas which are beyond reach. He focuses on a business which may be snubbed by others as trivial — condiments on dining tables — but he always holds the conviction that so long as we work hard, a brighter future will prevail. Such a simple but powerful conviction has prompted him to immerse in Lee Kum Kee wholeheartedly, and for decades, lead the company towards its goal with unyielding perseverance. That was by no means a smooth process, but was often beset with internal strife and external difficulties, and at times, in extremely precarious situations. However, Lee Man Tat neither retreated nor was intimidated by adversity and hindrance. He constantly responded with an indomitable spirit derived from his iron will, thus incessantly advancing Lee Kum Kee from a humble family business to a multinational corporation today.

For all of his stature, Lee Man Tat is modest and unadorned. He is also sincere and amiable in his interactions with others. Such an admirably unassuming gesture is certainly a fine model for the younger generations to follow.

The establishment of Hong Kong as an entrepôt in the 1840s under the British colonial rule marked the decline of China's supremacy and the beginning of Western acts of intimidation and humiliation. Many Chinese fell into desperate plights, and were therefore compelled to leave their hometowns for a better livelihood abroad. However, the force of such circumstances has unexpectedly created a huge network of overseas Chinese communities worldwide, and their caring and nostalgia towards their hometowns and motherland have created an impressive flow of remittances and a substantial oversea Chinese market. The popularity of Lee Kum Kee's oyster sauce amongst the overseas Chinese could first and foremost be attributed to its premium quality and good marketing, yet another equally important factor is the "cultural taste" closely associated with their hometown nostalgia. Hong Kong's prominent position as an international trade centre — China-backed and open to global — has also greatly facilitated Lee Kum Kee's subtle bridging role to connect most effectively the Chinese people at home and abroad.

After passing through years of vicissitudes, by the end of the 20th century, China has resumed its exercise of sovereignty over Hong Kong and Macau. One after another, myriads of overseas Chinese flocked to re-join the crusade for national rejuvenation. But before then, Lee Man Tat already had the foresight to invest heavily in the Mainland since the 1980s, contributing to the cause of national rejuvenation as well as tapping into the powerful momentum of the rise of China to further expand the international market of Lee Kum Kee. The developmental trajectory of Lee Kum Kee and Lee Man Tat vividly reflects what the Chinese communities in Hong Kong, Macau, and overseas have commonly experienced throughout the modern history of China. They have contributed momentously to the nation as well as witnessed the tortuous course of national rejuvenation from feeble to strong.

Precisely for the foregoing reasons Lee Man Tat's story is worth telling. We narrated the extraordinary experiences of his family, his business, and the industry; in a broader sense, we also narrated similar experiences of the contemporary Chinese in Hong Kong, Macau, and overseas communities, as this was a tortuous passage commonly shared by them throughout the history of modern China. The story started with a wok of oyster sauce to producing millions of bottles of sauces per day, and how it morphed into a multinational corporation from a secluded corner in South China to the world of overseas Chinese and eventually returned to China is truly meaningful and inspiring. To this end, we strove to track Lee Man Tat's footprints to be as detailed as possible, with the hope to possibly enlighten future generations the essence of this outstanding octogenarian, helping them to become the best they could be, beyond the histories of emperors and dignitaries, with a supplement of remarkable stories from ordinary people.

LEE MAN TAT:
CHRONICLE OF EVENTS

DATE	EVENT
1929.01.05	Born at No.11, Rua da Praia do Manduco, the residence of the Lee family in Macau; then named "Lee Kin Tai".
1932	Second generation of the Lee Kum Kee family relocated the company's headquarters from Macau to Hong Kong.
1932	The new Lee Kum Kee workshop-cum-store was located at No. 262, Queen's Road Central, Hong Kong.
1937	Enrolled in Chongshi School, Macau.
1941	Transferred to Pui Ying School, Macau, at Primary Five. Name changed to "Lee Man Tat".
1943	Continued his secondary education at Kwong Tai Middle School, Macau.
1946	Transferred to Chee Yung High School, Macau, for his senior secondary education, but soon left the school.
1946	Moved to Guangzhou. Managed Lee Kum Kee's properties in Guangzhou as well as making and selling oyster sauce there as a side-business.
1949	Left Guangzhou and returned to Macau.
1950	Established Tat Sang Company in Macau.
1950	Outbreak of Korean War led to trade embargo on China. The US-led embargo prohibited any Hong Kong product which adopted raw materials from Mainland China to be exported to the United States.
1952	Set up Man Sang Leatherware Factory in Macau. Expanded to a total of six factories thereafter.
1954.12.23	Married Choi May Ling. Officially joined Lee Kum Kee and moved to Hong Kong.
1956	Eddy Lee, his eldest son, born in Hong Kong.
1957	Elizabeth Lee, his eldest daughter, born in Hong Kong.
1958	David Lee, his second son, born in Hong Kong.
1960	Lee May Mui, his second daughter, born in Hong Kong.

1960	Closed Man Sang Leatherware Factory and other factories in Macau.
1961	Charlie Lee, his third son, born in Hong Kong.
1964	Sammy Lee, his fourth son, born in Hong Kong.
1965	Lee Kum Kee's licensee changed from Lee Man Kwong to Lee Man Tat after the "crisis of oyster extracts".
1960s	Received four scrolls of calligraphy from Liu Tai-shi as a gift in Taipei, and was particularly enlightened by the four characters of "Si Li Ji Ren".
1968	The old building at No. 262, Queen's Road Central, Hong Kong, where Lee Kum Kee's store was located was listed as a "dangerous building". The company temporarily moved to Catchick Street, Sai Wan.
Late 1960s	Disputes with the other two Lee families sharing the Lee Kum Kee ownership over the future of the company.
1972	The First Family Separation. Bought the remaining shares of Lee Kum Kee from the two other families, and launched the economical Panda Brand Oyster Sauce in the United States shortly after taking the helm of Lee Kum Kee.
1976	The new plant at Wong Chuk Hang, Hong Kong, commenced production. Steam boilers were introduced.
1976	Mother Chan Choi Kam died.
1979	Took part in the first voyage for foreign ships to go upstream along the Yangtze River to Central China.
1982	Relocated Lee Kum Kee to a new plant built in Tin Wan, Hong Kong.
1983	Set up the office of Lee Kum Kee (USA) Inc. in Los Angeles.
1984	Younger brother Lee Man Lok raised the issue of withdrawing from the family business to realise his share-holding.
1984	Lee Kum Kee filed lawsuit against a Malaysian factory and supermarkets concerned with producing and selling counterfeit Panda Brand Oyster Sauce.

1984.12.19	The signing of Sino-British Joint Declaration in Beijing. The British would hand over Hong Kong to China with effect from July 1, 1997.
1985	Suffered from acute colitis and underwent two surgeries in Los Angeles. Spent four months for recuperation in Los Angeles.
1985-1986	Lee Man Lok took legal action in court to seek a winding-up of Lee Kum Kee.
1986.04	Second Family Separation. Lee Man Lok and Lee Man Tat reached a settlement with Lee Man Tat buying Lee Man Lok's share of the business.
1986	Set up the second office of Lee Kum Kee (USA) Inc. in New York.
1987	Father Lee Shiu Nan died.
1988	Relocated Lee Kum Kee's headquarters and plant to Tai Po Industrial Estate, Hong Kong, to commence operation.
1990.10	The school donated by Lee Man Tat to commemorate his father, Lee Shiu Nan Memorial Primary School, was opened in Nanshui, Zhuhai.
1990.10	First trip to return to his ancestral hometown of Qibao, Xinhui.
1991	New plant in Los Angeles, USA, commenced production.
Early 1990s	Lee Kum Kee won the trademark infringement lawsuit in Malaysia. Lee Man Tat, however, decided not to proceed with the claims of compensation against the 18 local supermarkets.
1992.08.08	First met Professor Zhao Yunhong, the then President of the PLA First Military Medical University, in Guangzhou.
1992	Established Nanfang Lee Kum Kee Health Products Co. Ltd. as a joint venture with the PLA First Military Medical University in Guangzhou.
1992	Developed and launched Lee Kum Kee's XO Sauce.

1993	Bought Chue Lung Factory in Kwu Tung, Sheung Shui, Hong Kong, as a base to experiment with the soy sauce production.
1995	The new plant in Fuzhou, Fujian Province, commenced production, marking the starting point of Lee Kum Kee's investment in China.
1995	Decided to build a major production base in his ancestral hometown of Qibao, Xinhui.
1995	Lee Kum Kee became the first food manufacturer in Hong Kong to obtain the ISO 9002 certification.
1996	Lee Kum Kee Group's Xinhui production base commenced operations.
1997	New plant in Kuala Lumpur, Malaysia, commenced production.
1997	Developed and launched Lee Kum Kee's Seasoned Soy Sauce for seafood.
1997-1998	After the Asian Financial Crisis, Lee Man Tat decided to halt Lee Kum Kee's diversification in business and re-focused on its core business.
1998	New plant in Huangpu, Guangzhou, commenced operation.
1998	Opening of Lee Man Tat Secondary School in Qibao, Xinhui.
1998.04.21	Nanfang Lee Kum Kee seriously hit by the abrupt government policy change to ban any form of direct selling in Mainland China.
1998	Lee Man Tat purchased the remaining shares of Nanfang Lee Kum Kee from the PLA First Military Medical University after the State Council banned the military and all its associated units from engaging in commercial activities. The company became wholly owned by the Lee Kum Kee Group.

2000	Lee Man Tat decided to surmount the financial difficulties in the operation of Nanfang Lee Kum Kee and continued running the business through imposing strict budgetary measures to curb expenditure.
2002	Establishment of the "Family Council".
2004	Awarded the "Hong Kong Business Mastermind" by the Hong Kong General Chamber of Commerce.
2006.07.07	Installation as an Advisory Professor of Tsinghua University, and the opening of the Lee Man Tat Bio-Medical Science Library in the University.
2008	Lee Kum Kee became the official F&B supplier to the Beijing 2008 Olympics.
2008.04.09	Inducted into the Academy of Distinguished Entrepreneurs, Babson College, USA, as the first Chinese entrepreneur to receive such an award.
2008	Lee Kum Kee earned the "International Union of Food Science and Technology–Food Service Award" jointly presented by the International Union of Science and Technology and the Chinese Institute of Food Science and Technology.
2009	Renamed the Nanfang Lee Kum Kee as "Infinitus (China) Co. Ltd." to effectively differentiate the two distinctive brands of health products and sauce products.
2009	Received the "Industrialist of the Year Award 2009" by the Federation of Hong Kong Industries.
2010	Lee Kum Kee became the official qualified suppliers of raw and auxiliary materials in the catering category at Expo 2010, Shanghai.
2010	Awarded the Bronze Bauhinia Star by the Government of the Hong Kong Special Administrative Region.
2011	Made donations to build the "Lee Man Tat Bridge" and the "Infinitus Bridge" in Xinhui, Guangdong, to ease traffic congestion there.

2012	Lee Kum Kee became the "Official Partner of the China Space Industry". Since then, its products have followed the astronauts of Shenzhou IX (2012), X (2013) and XI (2016) into outer space as part of their food condiments.
2012	Received the "2012 Southern – Top Ten Chinese Philanthropists" by the Overseas Chinese Affairs Office of the People's Government of Guangdong Province.
2013	Coutts awarded the "Coutts Prize for Family Business Hong Kong 2013 - Family Governance" to Lee Kum Kee.
2013	Elected "Leader of the Year 2012" by the Sing Tao Group in the Commerce & Industry/Finance Category.
2014.11	Opening of the Lee Man Tat Bridge in Xinhui, Guangdong Province.
2016	Received the "Outstanding Achievement Award" by the Hong Kong Food Council.
2017.09	Opening of the Infinitus Bridge in Xinhui, Guangdong Province.
2017	Elected as the "Global Top Ten Outstanding Cantonese" at the Third Global Conference of the Cantonese.
2018	Awarded the Silver Bauhinia Star by the Government of the Hong Kong Special Administrative Region.
2018	Granted the Lifetime Achievement Award in the Hong Kong Business Awards 2018 (organized by DHL and the South China Morning Post) for his sustained business and social contributions to Hong Kong.

REFERENCES

Booth, M. 1996. *Opium: A History*. London: Simon & Schuster.

Cheung, G.K. 2009. *Hong Kong's Watershed: The 1967 Riots*. Hong Kong: Hong Kong University Press.

Clarence-Smith, W.G. 1985. *The Third Portuguese Empire, 1825-1975: A Study in Economic Imperialism*. Manchester, UK: Manchester University Press.

Gersick, K.E. (et al.) 1997. *Generation to Generation: Life Cycles of the Family Business*. Boston, Mass.: Harvard Business School Press.

Hicks, G. 1993. *Overseas Chinese Remittances from Southeast Asia: 1910-1940*. Singapore: Selected Books.

Lee, S. 2016. *The Autopilot Leadership Model* (Translated by Jamie Lee from《自動波領導模式》, 2012). Singapore: McGraw-Hill Education (Asia).

Ning, X. 2018. *Family Spirit: The Power behind the Inheritance of Lee Kum Kee through the Centuries*. Hong Kong: ET Press.

Poston, D.L. Jr. & Wong, J.H. 2016. "The Chinese diaspora: The current distribution of the overseas Chinese population", *Chinese Journal of Sociology*, vol. 2, no. 3, pp. 348-373.

Sanaa A. 2008. *IFC Family Business Governance Handbook*. Washington, DC: International Finance Corporation.

Schumpeter, J.A. 1934. *The Theory of Economic Development: An Inquiry into Profits, Capital, Credit, Interest, and the Business Cycle*. Cambridge: Harvard University.

Wang, G. 2003. *Anglo-Chinese Encounters since 1800: War, Trade, Science, and Governance*. Cambridge: Cambridge University Press.

Yim, B. & Mak, V. 2005. "Lee Kum Kee: Old company, modern marketing strategy", in Ali F. Farhoomand (ed.), *Small Business Management and Entrepreneurship in Hong Kong: A Casebook*, pp. 63-75. Hong Kong: Hong Kong University Press.

Chang, Qing 常青。1999。《百年澳門》(*One Hundred Years of Macau*)。北京：作家出版社。

Chen, Xiaobin 陳曉彬。1999。《澳門今昔》(*Macau Past and Present*)。北京：中國書籍出版社。

Deng, Kaisong; Lu, Xiaomin & Yang, Renfei 鄧開頌、陸曉敏、楊仁飛。2011。《澳門史話》(*Historical Stories of Macau*)。北京：社會科學文獻出版社。

Deng, Xiaoping 鄧小平。1993。《鄧小平文選》第三卷 (*Selected Works of Deng Xiaoping, Volume III*)。北京：人民出版社。

Feng, Bangyan 馮邦彥。1999。《澳門概論》(*An Introduction to Macau*)。香港：三聯書店（香港）有限公司。

Fu, Lianwan 傅連婉。2012。〈廣搜蘭蕙慰桑榆——記信豐才子劉太希〉(On Liu Tai-shi, a notable talent from Xinfeng)，原出處不詳，2014 年 8 月 12 日刊於《客家搖籃網》(*Hakka Cradle*)，http://

www.gnhakka.com/n492/c7489/content.html

Guo, Tingyi (Kuo, Ting-yee) 郭 廷 以。1979。《近代中國史綱》（上下冊）(*A Short History of Modern China, Volume I & II*)。香港：中文大學出版社。

Huaqiao Ribao (Wah Kiu Yat Po) 華僑日報。1948。《1948 年香港年鑑》(*Hong Kong Annal 1948*)。香港：華僑日報有限公司。

Huang, Qichen 黃啓臣。1999。《澳門通史》(*A General History of Macau*)。廣州：廣東教育出版社。

Jiang, Shunyuan 姜舜源。2012。〈張大千「澳門時代」墨寶面世〉(Paintings painted by Chang Dai-chien in Macau resurface)，《澳門雜誌》(*Macauzine*) 總第 87 期，頁 67-73。

Lao, Jiayu (Lou Ka U) 勞 加 裕。2014.06。〈濠江‧蠔情——百年澳門蠔油業滋味〉(A taste of it: Hundred years of oyster sauce industry in Macau)，《梳打雜誌》(*SODA Magazine*) 第 63 期，頁 66-71。

Lei, Ge 雷戈。2012。《因獻成文：中國大陸官修野史研究‧第十一章：統戰實踐》，轉引自〈中共利用文獻資料統戰港臺海外和西方〉(Literature as a tool for "United Front" campaign: Strategy of Chinese Communist Party towards Overseas Chinese in Hong Kong, Taiwan, and the West)，《黃花崗雜誌》(*Huang Hua Gang Magazine*) 總第 41 期，頁 23-56。

Li, Fulin 李福麟。2009。《澳門舊話》(*Historical Talks of Macau*)。澳門：澳門松山學會。

Li, Huisen (Lee, Wai Sum) 李惠森。2012。《思利及人的力量：成就一生的 9 個法則》(*The Power of "Si Li Ji Ren": Nine Rules for Life Achievement*)。北京：中信出版社。

Liang, Zhangju 梁章鉅。1987。《楹聯叢話》(*Notes on Couplet*)。北京：中華書局。

Liang, Zhenxing 梁振興。1990。〈李錦裳與「李錦記蠔油」〉(Lee Kum Sheung and the Lee Kum Kee Oyster Sauce)，《珠海文史》(*Humanity and History of Zhuhai*) 第 9 輯，頁 94-100。珠海：珠海市政協文史資料委員會。

Ma, Yue (Ma, Ngok) 馬 嶽。2010。《香港政治：發展歷程與核心課題》(*Hong Kong Politics: Development Process and Key Issues*)。香港：香港中文大學香港亞太研究所。

Nanfang Ribao《南方日報》。2002。〈第一軍醫大學原校長趙雲宏同志逝世〉(Zhao Yunhong, former President of the First Military Medical University, has died)，《南方日報》，2002 年 11 月 19 日。

Situ, Biao & Tang, Shijin 司徒標、唐仕進。1995。〈文史牽動赤子情——珠海市政協通過文史工作聯絡「三胞」片段〉(Zhuhai Political Consultative reached out to Chinese worldwide beyond Mainland China through works on literature)，《文史通訊》(*Newsletter on Humanity and History*)，1995 年第 2 期。

Sun, Feng 孫鋒。2014。〈海天調位戰〉(Commercial wars between Lee Kum Kee and Haitian)，《商界》(*Business*) 總第 487 期，頁 64-67。

Wen, Dingming 溫鼎銘。1990。〈我所認識的蔡克庭先生〉(Choi Hac Teng: The gentleman that I knew),《珠海文史》(*Humanity and History of Zhuhai*) 第 9 輯,頁 101-103。珠海:珠海市政協文史資料委員會。

Wu, Yunpu 武雲溥。2013。〈李錦記:香港蠔油大王傳奇〉(Lee Kum Kee: The legend of Hong Kong's oyster sauce king),《國家人文歷史》(*National Humanity History*),2013 年第 2 期,總第 74 期,頁 81-85。北京:人民日報社。

Yazhou Dianshi Xinwenbu Zixunke 亞洲電視新聞部資訊科。2009。《解密五百年澳門》(*Five Hundred Years of Macau: An Anatomy*)。香港:明報出版社。

Ye, Lingfeng (Yeh, Ling-feng) 葉靈鳳。1999。〈蠔和蠔田〉(Oyster and oyster farming),刊於《靈魂的歸來》(葉靈鳳文集第二卷:散文、小品) (*Return of the Soul: Essay Collections of Yeh Ling-feng*), 頁 352-355。廣州:花城出版社。

Xin Bao (Hong Kong Economic Journal)《信報》。2017。〈中港百年家族企業 293 間・僅及日本 1%〉(A total of 293 century-old family enterprises in mainland China and Hong Kong — only 1% of Japan's number),《信報》,2017 年 9 月 13 日。

Yi Zhoukan (Next Magazine)《壹週刊》。2007。〈百億海產大王蔡繼有傳奇〉(Seafood multi-billionaire: The legend of Choi Kai Yau),《壹週刊》第 921 期,頁 54-59。

Yi Zhoukan (Next Magazine)《壹週刊》。2014。〈百年蠔門李錦記〉(Lee Kum Kee: Hundred years of oyster sauce business),(台灣)《壹週刊》第

676 期,頁 76-79。

Yuan, Qiushi 袁求實。1997。《香港回歸大事記:1979-1997》(*Hong Kong's Transition 1979-1997: Chronology and Selected Documents*)。香港:三聯書店(香港)有限公司。

Zhang, Jiawei (Cheung, Ka Wai) 張家偉。2000。《香港六七暴動內情》(*Inside Stories of the 1967 Riots*)。香港:太平洋世紀出版社。

Zhang, Yinglong (ed.) 張英龍(主編)。2009。《珠海僑務誌》(*Overseas Chinese Affairs of Zhuhai*)。珠海:珠海出版社。

Zheng, Baohong (Cheng, Po Hung) 鄭寶鴻。2016。〈第五章:金山莊與其他莊行〉(Chapter 5: Kam Shan Chong and other trading agents),《香港華洋行業百年:貿易與金融篇》(*Trade and Financial Industries in Hong Kong Over a Century*),頁 50-53。香港:商務印書館。

Zheng, Baohong (Cheng, Po Hung) 鄭寶鴻。2016。《香港華洋行業百年——飲食與娛樂篇》(*Food and Entertainment Industries in Hong Kong Over a Century*)。香港:商務印書館。

Zheng, Hongtai & Huang, Shaolun (Zheng, Wan Tai & Wong, Siu Lun) 鄭宏泰、黃紹倫。2006。《香港股史 1941-1997》(*History of Hong Kong Stock Markets, 1941-1997*)。香港:三聯書店(香港)有限公司。

Zheng, Hongtai & Lu, Guanhao (Zheng, Wan Tai & Luk Koon Hoo) 鄭宏泰、陸觀豪。2017。《點石成金:打造香港金融中心的里程碑》(*The Midas Touch: Milestones in Shaping Hong Kong as a*

Financial Centre)。香港：中華書局。

Zhuhaishi Difangzhi Bianweihui (ed.) 珠海市地方
誌編委會（編）。2001。《珠海市誌》(*Chronicle
of Zhuhai City*)。珠海：珠海市地方誌辦公室。

Zhuang, Guotu 莊 國 土。2009。〈 東 南 亞 華
僑華人數量的新估算 〉(New estimation on the
population of Southeast Asian Chinese)，《 廈 門
大學學報》（哲學社會科學版），2009 年第 3 期，
總第 193 期，頁 62-69。

Capital Entrepreneur《資本企業家》。2014。〈李
錦記第四代傳人 • 李惠中：沒有守業，只有永遠
創業〉(Charlie Lee, fourth generation of Lee Kum
Kee: We abide by constant entrepreneurship, not
simply keeping it)，《資本企業家》第 113 期，頁
22-27。

Zhong, Shiyuan (Chung Sze Yuen)   鍾 士 元。
2001。《香港回歸歷程：鍾士元回憶錄》(*Hong
Kong's Journey to Reunification: Memoirs of Sze-yuen
Chung*)。香港：中文大學出版社。

Zou, Guangwen (ed.) 鄒廣文（主編）。2010。《民
族企業品牌之路：李錦記集團發展歷程分析》
(*Towards a Prestigious Brand of National Enterprise:
An Analysis of the Development of Lee Kum Kee
Group*)。香港：經濟日報出版社。

三聯書店
http://jointpublishing.com

JPBooks.Plus
http://jpbooks.plus

BIOGRAPHY OF LEE MAN TAT - Legend of the Sauce King

Author   Kok Chung ONG, Victor ZHENG, Siu-lun WONG
Editor   Carrie Kwai, Ling Cho Fung
First Published in January 2019

Published by Joint Publishing (H.K.) Co., Ltd.
20/F., North Point Industrial Building, 499 King's Road, North Point, Hong Kong

Printed by Elegance Printing & Book Binding Co., Ltd.
Block A, 4/F., 6 Wing Yip Street, Kwun Tong, Kowloon, Hong Kong
Distributed by SUP Publishing Logistics (HK) Ltd.
3/F., 36 Ting Lai Road, Tai Po, N.T., Hong Kong

Copyright © 2019 Joint Publishing (H.K.) Co., Ltd.
Published & Printed in Hong Kong
ISBN 978-962-04-4428-9